The Deliberate Corruption of Climate Science

Tim Ball, PhD

The Deliberate Corruption of Climate Science

ISBN 978-0-9888777-4-0

eBook ISBN 978-0-9888777-5-7

STAIRWAY≡PRESS

STAIRWAY PRESS—SEATTLE

Cover Design by Josh
http://www.cartoonsbyjosh.com/

www.StairwayPress.com
1500A East College Way #554
Mount Vernon, WA 98273

Acknowledgements

The price paid for a person's activities is always higher for those who support them. They are the innocent bystanders, the ones without who individuals could not achieve anything. I am eternally grateful for my family support, especially wife Marty, sons Dave and Tim Jr. and sister Liz.

Voltaire said, "It is dangerous to be right in matters where established men are wrong." The danger extends to those who support the person who challenges "established" men. I am grateful for all those who provided moral, social, philosophical, intellectual and financial support.

I appreciate the patience, wisdom, experience and skill provided by the publisher Ken Coffman and my editor Stacey Benson.

—Tim

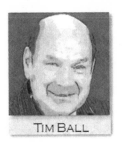

TIM BALL

PREFACE

Then up and spent the evening walking with my wife, talking; and it thundering and lightening mightily all the evening—and this year have had the most thunder and lightening, they say, of any in man's memory.
—Diary of Samuel Pepys, entry for July 3rd, 1664

I'VE STUDIED CLIMATE both scientifically and academically for over forty years after spending eight years studying meteorology and observing the weather as an aircrew and operations officer in the Canadian Air Force. When I began the academic portion of my career, global cooling was the concern, but it was not a major social theme. During the 1980s the concern switched to global warming which became a major political, social and economic issue.

I watched my chosen discipline—climatology—get hijacked and exploited in service of a political agenda, watched people who knew little or nothing enter the fray and watched scientists become involved for political or funding reasons—willing to corrupt the science, or, at least, ignore what was really going on. The tale is more than a sad story because it set climatology back thirty years and damaged the credibility of science in general.

It also undermined the environmental movement by

incorrectly claiming massive environmental damage and setting up a classic 'cry wolf' scenario. It is the greatest deception in history and the extent of the damage has yet to be exposed and measured.

There have been, of course, other sad deceptions throughout history, but all of them were regional, or, at most, continental. The deceptive idea that human-generated CO_2 causes global warming or climate change impacted every person in the entire world, thus it reflects Marshall McLuhan's concept of the *global village*. This book shows how the deception was designed to be global by involving every nation through the agencies of the United Nations. Historians with the benefit of 20:20 hindsight will wonder how such a small group was able to achieve such a massive deception. There are several reasons why the public was deceived:

1. The objective and therefore the science were premeditated.
2. The scientific focus was deliberately narrowed to CO_2.
3. From the start, unaccountable government agencies were involved and in control.
4. Science and political structures and procedures were put in place to enhance the deception.
5. Actions were taken to block or divert challenges.
6. The people's natural fears about change and catastrophe were exploited.
7. The public's lack of scientific understanding, especially with regard to climate science, was exploited.
8. People find it hard to believe a deception on such a grand scale could occur.
9. Opponents were ruthlessly attacked, causing others to remain silent.

Preface

Some call the human-caused global warming theme a hoax, but that's incorrect: a hoax is defined as a humorous or malicious deception. The Piltdown Man[1] was a hoax perpetrated by one academic to expose the arrogance and pomposity of another. Its impact was in academia, but had little relevance in the real world. There is nothing humorous about the corruption of climate science. Further, a political objective need not be malevolent; however, the methods used to achieve the goals of progressive activists are assuredly ugly, malicious and wrong.

Some have called the corruption of climate science a conspiracy partly because conspiratorial themes are fueled by speculation on the Internet. A conspiracy is defined as a secret plan to do something unlawful or harmful. There is no doubt what the activists have done is harmful, but pursuing a political goal is lawful. What is unlawful is using deliberate deceptions, misinformation, manipulation of records and misapplying scientific method and research. Indeed, it is amazing how they deceived the entire world through using existing laws and societal structures; it fits the classic description of daylight robbery.

It is more appropriate to identify the group as a cabal: a secret political clique or faction. This book explains their motive and objectives which were political, not scientific. It explains how in order to do this they bypassed and perverted the scientific method—the normal and proper method by which science

---

[1] Even today the name 'Piltdown' sends a shiver through the scientific community, for this quiet Sussex village was the site of a dramatic and daring fraud, the fallout from which continues to affect us.

Between 1908 and 1912, the discovery of human skull fragments, an ape-like jaw and crudely worked flints close to Piltdown was hailed by the world's press as the most sensational archaeological find ever: the 'missing link' that conclusively proved Charles Darwin's theory of evolution.

The Piltdown Man Hoax: Case Closed, Miles Russell, The History Press

progresses. They effectively silenced scientists who tried to perform the normal roles of critics and skeptics.

Consider this brave but late (May 12th, 2012) admission by German Physicist and meteorologist Klaus-Eckart Plus:

> Ten years ago I simply parroted what the IPCC told us. One day I started checking the facts and data—first I started with a sense of doubt but then I became outraged when I discovered that much of what the IPCC and the media were telling us was sheer nonsense and was not even supported by any scientific facts and measurements. To this day I still feel shame that as a scientist I made presentations of their science without first checking it.
>
> ...scientifically it is sheer absurdity to think we can get a nice climate by turning a CO_2 adjustment knob.[2]

If someone so knowledgeable about the subjects of meteorology and atmospheric physics can be so readily deceived, it is not surprising that the general public was deceived. This underscores the effectiveness of the deliberate and carefully orchestrated climate science deception.

However, it also underscores the problems of writing a book that identifies what they did and how it was done. If the science is too complicated, people won't read the book. If it is too simplistic, the critics will say it trivializes and makes errors that they will use to condemn. As I hope you will learn, the architects of this greatest deception actually set up a web site to carry out such attacks on those who dared to question. They recruited the mainstream media to run stories marginalizing the scientific claims

---

[2] http://notrickszone.com/2012/05/09/the-belief-that-co2-can-regulate-climate-is-sheer-absurdity-says-prominent-german-meteorologist/

and the credibility of the messenger. Seth Borenstein, a national science writer for the Associate Press, sent an email on July 23rd, 2009 to the Climatic Research Unit (CRU) gang whose corruptive practices were exposed by leaked emails. He wrote:

> *Kevin, Gavin, Mike, It's Seth again. Attached is a paper in JGR today that Marc Morano is hyping wildly. It's in a legit journal. Watchya think?*[3]

A journalist talking to scientists is legitimate, but like the leaked emails, the tone and subjective comments are telling. *"Again"* means there is previous communication. Others commented on Borenstein being *too damn cozy with the people he covers.*[4] Another example was the unhealthy connection between Richard Black of the British Broadcasting Corporation (BBC) and the CRU. As Michael Mann wrote:

> *...extremely disappointing to see something like this appear on BBC. its particularly odd, since climate is usually Richard Black's beat at BBC (and he does a great job).*
>
> *We may do something about this on RealClimate, but meanwhile it might be appropriate for the Met Office to have a say about this, I might ask Richard Black what's up here?*[5]

---

[3] http://m4gw.com/agw-proponents-fight-rearguard-action-as-political-climate-science-fails/

[4] http://wattsupwiththat.com/2009/12/12/aps-seth-borenstein-is-just-too-damn-cozy-with-the-people-he-covers-time-for-ap-to-do-somethig-about-it/

[5] http://toryaardvark.com/2009/12/27/climategate-the-bbcs-unhealthy-connection-with-michael-mann/

These are just a fraction of the machinations that went on to achieve their goal of bringing down industrialized economies and societies using science.

I broke with recommended procedures and used many quotes because it is necessary to show they knew what they were doing—that is it was premeditated. I reduce science to a minimum and only show major misuse, not to mislead, although I will likely be charged with "cherry-picking", but to illustrate and support the overall story.

This is not intended as a textbook, nor is it meant to be an academic treatise, although it has many footnotes. It is an attempt to explain what happens when science is misused for a political agenda and this is only achieved by creating a hybrid book form. This approach is required because the world has never before suffered from deception on such a scale.

RICHARD LINDZEN

CHAPTER ONE

Historical Development

MOST PEOPLE HAVE no idea that climate changes—most are unaware of the extent and speed with which climate changes and that is a vital part of the problem with the current state of climate science.

Climate change has happened, is happening and will always happen. Contrary to the message of the last thirty years, current rate of climate change is well within the bounds of natural variability. Thus, a perfectly natural phenomenon became the biggest deception in history.

How was it done? What was the motive?

Figure 1 shows a simple systems diagram of weather components and interactions. Given all these variables, we must ask why only one variable—carbon dioxide (CO_2)—became the focus of climate science, a global environmental movement and transnational energy policy.

CO_2 is not visible in the diagram because it is one minuscule part of the atmospheric influences on climate. It would be under "Atmospheric Composition" in center right of the diagram. Why and how it became so important is the key to understanding what has gone on over the last thirty years.

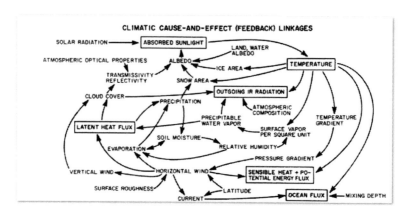

Figure 1

Climatic Cause-and-Effect (Feedback) Linkages[6]

Phrases such as global warming, the greenhouse effect and climate change are part of everyday language, but few people truly understand the details. Even folks who supported and promoted these concepts and the policies they engendered don't necessarily understand them. Richard Lindzen, atmospheric physicist, and Alfred P. Sloan, Professor of Meteorology at the Massachusetts Institute of Technology, said in 1989:

> *The issue of the 'greenhouse effect' has assumed a peculiar life of its own. Politicians, government officials, and various policy specialists cling with remarkable tenacity to the notion that this is a proven and intolerable danger about which there is scientific*

---

[6] *Climate Stabilization: For Better or for Worse?* William W. Kellogg and Stephen H. Schneider, Science, Volume 186, December 27, 1974 http://www.sciencemag.org/content/186/4170/1163.full.pdf?ijkey= 9f44624ad3ffeba22871a30e3d100d92d8bbd4c7&keytype2=tf_ipsecsha

*unanimity. At the same time, one has no difficulty
hearing the muttering in the corridors of any meteorology
department that this is an issue that has gotten out of
hand, that the claims are insupportable, that the models
are inadequate, and the data contradictory.*[7]

How did the situation Lindzen described happened? A major
reason is because Climatology is the study of weather averages in a
region or over time, which includes all the components of the
diagram and more. However, Climate Science involves specialists
who only consider one component of the diagram, often in
isolation from the other components. It fits the adage that they
can't see the forest for the trees.

Current climate changes are natural, so why do a group of
influential people suggest they are not? Many believe the current
state of climate change is unnatural, unique and unprecedented.
To answer that question requires a brief historical and
philosophical overview for context. We don't just suddenly arrive
at situations unless it is pure catastrophe. There is always a
history, and the current situation can be understood when it is
placed in context.

More people are starting to understand that what they're
told about climate change by academia, the mass media and the
government is wrong, especially the propaganda coming from the
UN and the Intergovernmental Panel on Climate Change (IPCC).
Ridiculous claims—like *the science is settled* or *the debate is over*—
triggered a growing realization that something was wrong. There
were attempts to explain the contradiction between what people
were told and what they experienced. It led to comments so
illogical that even those who didn't understand the science
recognized the illogic. Cooling is due to warming? Cartoonists are
masters at exploiting such illogic.

---

[7] As reported by John L. Daly, http://www.john-daly.com/quotes.htm

Actually, this pattern shows how societal views and beliefs change. Extremists and extremism define the limits for the majority. Extremism will increase as purveyors of prevailing views on climate and climate change try to defend the indefensible.

Realization of the misdirection, and in too many cases, the deception, leads to the next stage of logical questions. What is the motive? How did the situation develop? How did society become so easily misled? Why do some, including governments, continue to accept or push the prevailing wisdom, even when it's clearly wrong? In this and subsequent chapters, I will provide answers. Like all prevailing philosophies, these themes are a part of the evolution of ideas and a historical context is useful.

Changing Western Views of the World

The current western view of the natural world essentially evolved from the work of Charles Darwin. Though evolutionary theory is not proven law after 150 years of effort, it is now the only view allowed in most public schools.

Why?

Not allowing other ideas to be presented and discussed suggests there is fear of the other ideas—it is a measure of insecurity. It parallels the fear of free speech inherent in totalitarian regimes. It is similar to the consensus thinking behind the claim that climate science is settled. One myth created by this thinking is that current science is the truth. Actually, it is only today's truth that replaced yesterday's truth and will be replaced by tomorrow's truth.

The age of the earth as determined by western society is essential to this discussion. It is also necessary to understand the larger philosophy underpinning western science, of which climate change is a subset. Christianity said God created the world in seven days. The English church accepted Bishop Ussher's (1581-

1656 A.D.) biblically based calculation that God created the world on Sunday, October 23rd, 4004 BC. Darwin needed a much older world with time to allow evolution to occur according to his theory. Darwin needed millions, although both agree it is a work in progress. Sir Charles Lyell provided a more suitable timescale for Darwin in his book *Principles of Geology*. Darwin recorded in his 1839 book *Journal and Remarks,* more commonly called *The Voyages of the Beagle,* that he took a copy on his famous visit to the Galapagos Islands.

The combination of a long time period and slow development resulted in a philosophical view originally known as Uniformitarianism's—now more commonly known as Gradualism. Uniformitarianism sounds like a religious belief system because it is. It's a belief system that underpins the western societal view of the world. It replaced Neptunism, the religious view of before and after the Flood.

A brief flirtation with chaos theory in the 1970s coincided with evidence that an asteroid had wiped out the dinosaurs 65 million years ago. Stephen Jay Gould proposed a compromise called *punctuated equilibrium* to accommodate catastrophic events like the asteroid, but with the concept of little change over time.

Regardless, the basic tenet that assumes change is gradual over long periods of time means that any sudden or dramatic change is unnatural. It is easy to see how most people today are vulnerable to this idea because the philosophy permeates most school lessons, popular literature and the mass media. Environmentalists can point at such a natural change and say it is unnatural, which implies it is manmade and therefore a problem. Notice the illogic. If what we are doing is *unnatural,* then *we* are unnatural. If we are unnatural, it forces the questions of who we are and why we are here.

One obvious answer is a Greater Being put us here, but the entire essence of Darwin's theory is: there is no God. Darwin was a professed atheist. This debate still rages, but the public are unaware of the connection—they see the theory of evolution as a purely religious issue. Actually, it is part of the entire question of environmentalism and the misdirection discussed in this book. It is a part of today's debate manifest in bestselling books like Richard Dawkins' *The God Delusion* or Christopher Hitchens' *God is not Great*. Dawkins talks about Darwin as if he is a god himself.

Historically, new views of the world take time to infiltrate and become integral to the generic fabric of society. Some never accept them. Over 400 years ago, Copernicus correctly discovered that the earth circles the sun, but a late 20th century European poll showed a percentage still believe the sun goes round the Earth.[8] Their truth is based on experience—we don't appear to move while the sun appears to move across the sky. On a larger scale, to many, the physical reality doesn't matter. As long as the sun appears to sweep across the sky, they couldn't care less that the opposite is the real situation.

Similarly, Newton's theory of gravity is of immense scientific consequence, but is unfathomable to most people and of no consequence, as long they don't fall off the Earth. Interestingly, in his masterly work *Principia Mathematica,* Newton attributed planetary motion to physics, but ascribed the force of gravity to God, even though he could predict its effects.

Darwin's theory had much greater implication for everybody. To use a flippant exaggeration, he effectively said there is no God and your great grandmother was a gorilla. Then he began talking about everybody in a personal way, which is why reaction to his theory was so much more visceral—it widened the gap between the apolitical, amoral approach of science and the emotional.

Darwin was aware of the limitations of his theory. Unfortunately, he, or at least his theory, became a weapon for

---

[8] 1999 Gallup poll, http://www.gallup.com/poll/3742/new-poll-gauges-americans-general-knowledge-levels.aspx

defeating religion. As a result, scientists who wanted to pursue the scientific method and challenge the theory faced a problem. Considerable reaction occurred after publication of *On the Origin of Species* in 1859, dominated by religious and social condemnation. Science was more guarded because it assumed the theory would be tested following the scientific method. It did not take long for it to become difficult for any scientist to challenge or even question. Thus, evolutionary theory became entrenched and has been violently defended ever since.

The foremost legal challenge that established Darwin's Theory as the basis of modern science education was the Scopes Trial in 1925.[9] Ironically, this case challenged the Tennessee legislation that banned the teaching of evolution. Now, we are in the position where it is not illegal to teach its opposite philosophy, creationism, but it is almost universally excluded. This puts modern science into the same position religion faced.

In defending its position and views, post-modern science has become as dogmatic as the religion it replaced.

---

[9] Scopes Trial, beginning July 10th, 1925, at Dayton, Tenn., and lasting 11 days, was one of the most widely publicized legal cases in 20th century U.S. history. The charge was that John Thomas Scopes, a teacher of science in Rhea high school in Dayton, had violated the Tennessee state law prohibiting the teaching in public schools of any theories that deny the divine creation of man as taught in the Bible. Scopes, a biologist, had been teaching evolution. Scopes was convicted and fined $100. The defense appealed the case to the state supreme court, which in 1927, upheld the constitutionality of the 1925 law, but cleared Scopes on a technicality.
Source: Encyclopedia Britannica, 1961, Vol. 20, p. 132-133, Harry Elmer Barnes.

MICHAEL CRICHTON

CHAPTER TWO

Paradigm Shifts: New Ways of Seeing Things

ACADEMICS TALK ABOUT paradigm shifts, defined as *a fundamental change in approach or underlying assumptions.* They occur all the time and tend to follow similar patterns of adoption. Two currently at different stages are feminism and environmentalism. Society adopts the new paradigm with a small percentage taking it up quickly. Another small group will never accept the paradigm, while the majority generally accept, but remain unsure how far they should go.

Environmentalism was a necessary paradigm shift that took shape and gained acceptance in western society in the 1960s. The idea that we shouldn't despoil our nest and must live within the limits of global resources is fundamental and self-evident. Every rational person embraces those concepts, but some took different approaches that brought us to where we are now. How far shall we take the idea of environmentalism in terms of restricting or punishing people's behavior?

Symbolic events are most effective in shifting our thinking. Pictures of the Earth from Apollo 8 taken in December 1968 triggered the paradigm shift of environmentalism. Suddenly, our planet seemed small and limited. Suddenly, it appeared there were too many people and we were running out of everything.

The Earth as Viewed from Space

Now let's examine how politics took over environmentalism and the subset climate change through exploitation of fear and ignorance. Michael Crichton's book *State of Fear* is a first class exploration of how special interest groups operate. The manipulators of environmentalism and climate exploited the public's lack of knowledge—presenting climate change as something new and occurring faster than natural.

Environmentalism: a Necessary, but Exploited Idea

A few took the title of environmentalist upon themselves when they have no qualifications—they appear on television with the subtitle *environmentalist* as if they are experts. The truth is almost everyone is an environmentalist and, therefore, qualifies. The question is how have a few usurped the moral high ground to claim that only they care about nature and our planet? My career involved working with farmers, foresters and rural people who are acutely aware of the environment and environmental issues because it is their backyard and their livelihoods.

Thomas Robert Malthus

The Club of Rome took the idea of environmentalism and adapted it to the work of Thomas Malthus (1766-1834). He assumed there was a direct relationship between population and food supply and predicted the population would outgrow our ability to nurture ourselves. As one group[10] commented:

> *Malthus was a political economist who was concerned about, what he saw as, the decline of living conditions in nineteenth century England. He blamed this decline on three elements: The overproduction of young; the inability of resources to keep up with the rising human population; and the irresponsibility of the lower classes. To combat this, Malthus suggested the family size of the lower class ought to be regulated such that poor families do not produce more children than they can support.*

This is important for the modern debate because overpopulation is still central to the use of climate change as a political vehicle.

Malthus had a crucial influence on the theory of evolution as Darwin acknowledged in his 1876 autobiography:

> *In October 1838, that is, fifteen months after I had begun my systematic inquiry, I happened to read for amusement Malthus on Population, and being well prepared to appreciate the struggle for existence which everywhere goes on from long-continued observation of the habits of animals and plants, it at once struck me that under these circumstances favourable variations would tend to be preserved, and unfavourable ones to be destroyed. The results of this would be the formation of a new species. Here, then I had at last got a theory by which to work.*

---

[10] Source: http://www.iisd.ca/Cairo/program/p03001.html

Malthus and Darwin ignored the benefits of technological advances, apparently because they were only interested in biological evolution. They didn't include in their thinking the British Agricultural Revolution that preceded the Industrial Revolution. This omission still pervades society today as many assume evolution has stopped. Another central, underlying theme of environmentalism is that technology is a dangerous anomaly in human development and underscores demands for sustainable development.

The most recent flurry of alarmism over population growth was central to the ideas of the Club of Rome, which received momentum through Paul Ehrlich's egregious and incorrect book, *The Population Bomb*. The fact that most predictions Ehrlich and John Holdren, advisor to President Obama for Science and Technology, made have proved wrong didn't stop extremists feeding their need for total control. Some believe people should not exist. Holdren thinks his fellow humans should be limited and controlled. Here is a totalitarian list of his proposals:[11]

1. *Women could be forced to abort their pregnancies, whether they wanted to or not.*
2. *The population at large could be sterilized by infertility drugs intentionally put into the nation's drinking water or in food.*
3. *Single mothers and teen mothers should have their babies seized from them against their will and given away to other couples to raise.*
4. *People who "contribute to social deterioration" (i.e. undesirables) "can be required by law to exercise reproductive responsibility"—in other words, be compelled to have abortions or be sterilized.*

---

[11] http://zombietime.com/john_holdren/

5. *A transnational "Planetary Regime" should assume control of the global economy and also dictate the most intimate details of Americans' lives—using an armed international police force.*

Challenges to these views occurred during his confirmation hearings for the White House position. He said they were no longer his views, but *saying* is easy. These assertions were so strongly held, it is unlikely they would change much. Can we identify an epiphany where his worldview reversed itself? We will examine his involvement in perverting climate science later—a sure indication that his thoughts are unchanged.

Anti-humanity in the Environmental Movement

Ron Arnold, Executive Vice-President of the Center for the Defense of Free Enterprise said:

> *Environmentalism intends to transform government, economy, and society in order to liberate nature from human exploitation.*

Mr. Arnold shines a spotlight on the illogic of the activists who undoubtedly believe in Darwinian evolutionary theory. Do the activists mean we are unnatural? Why isn't what has happened part of evolution? Of course, the extremists assume the virus won't begin with them. Prince Philip made the same assumption when he said:

> *In the event that I am reincarnated, I would like to return as a deadly virus, in order to contribute something to solve overpopulation...We need to 'cull' the surplus*

population.[12]

Maybe we should start with Monarchs and Royalty.

Ingrid Newkirk of People for the Ethical Treatment of Animals expressed similar ideas:

> *Mankind is a cancer; we're the biggest blight on the face of the earth.*[13]

Further...

> *If you haven't given voluntary human extinction much thought before, the idea of a world with no people in it may seem strange. But, if you give it a chance, I think you might agree that the extinction of Homo Sapiens would mean survival for millions if not billions, of Earth-dwelling species. Phasing out the human race will solve every problem on earth, social and environmental.*[14]

It is hard for most of us to get our minds around that. Here are Richard Conniff's comment in *Audubon*:

> *Among environmentalists sharing two or three beers, the notion is quite common that if only some calamity could wipe out the entire human race, other species might, once again have a chance.*[15]

---

[12] http://www.propagandamatrix.com/prince.html

[13] Quoted in *Beyond Cruelty,* Katie McCabe, Washingtonian, February 1990, p. 191.

[14] *Voluntary Human Extinction,* Les U. Knight, Wild Earth, Vol. 1, No. 2, (Summer 1991), p. 72.

[15] Richard Conniff, *Fuzzy-Wuzzy Thinking,* Audubon Society Magazine, November 1990.

Chapter Two

Elimination of humans is cynically and appropriately captured in the bumper sticker that says: *Save the planet, kill yourself.*

This extreme view is clearly nonsensical. In their opinion, even the people who hold the view that humans should not exist would include themselves and there is a philosophical and intellectual contradiction in this view and even the more extreme views of The Club of Rome. They accept without question the Darwinian view of nature and evolution. This holds that species evolve and adapt with the most successful surviving because of better adaptation. The phrase *"survival of the fittest"* was first used by English philosopher Herbert Spencer to describe his interpretation of Darwin's first edition of the *Origin of the Species.* He believed it supported his economic theories and thus underpinned the entire subject of what became known as Social Darwinism.

Darwin adopted the phrase "survival of the fittest" in the fifth edition of his book. However some claim he used it as an alternative for natural selection. Others argue that the concept of Social Darwinism is misapplied. The problem on the other side is somehow people think that evolution has stopped—this is another part of the impact of uniformitarianism. Even if they believe evolution is still occurring, it is so slow it would not be evident in the short span of a human life. If they accept Darwin's views, then humans are the most successful by definition. This is why Darwin considered human evolution separately in his book *The Descent of Man.*

Canadian David Suzuki, a former genetics professor and ardent environmentalist, apparently doesn't agree with Darwin. He said:

> *Economics is a very species—chauvinistic idea. No other species on earth—and there are maybe 30 million of them—has had the nerve to put forth a concept called economics, in which one species, us, declares the right to*

> *put value on everything else on earth, in the living and*
> *non-living world.*

First, he is wrong because all other species do put a value on everything else—it is food or it is not food. Doesn't get more basic than that. Second, 30 million is wrong, but so are the statistics Suzuki uses about the rate of extinctions. Besides, he really needs to think about what he said as it underscores the vast gap between humans and their abilities and *all* other plants and animals.

The Role of Extremists

Environmentalism was a good and necessary change, but like all new paradigms, a sequence of adoption and adaptation must follow. Initially people are wary because of our innate resistance to change. Gradually paradigms take hold as people realize their value. Environmentalism made us aware we had to live within the limits of our home and its resources: we had a responsibility for good stewardship. The problem is most people don't know how far to take a new paradigm or what good stewardship means.

For years, I wondered what extremists constructively add to a debate, but I've learned extremism usefully defines the limits of a new paradigm for the majority. By taking extreme positions, activists cause the majority to say, "Hold on, now you are going too far." School students respond with cheers and shouts when I ask them if they care about the environment. When I say, "Fine, then none of you will ever drive cars." I see a look on their faces that says, "Well, I don't care that much." I defined the limits of their environmentalism. As you would expect, it made them uncomfortable.

We've reached that point with environmentalism among the public. Others already realized the limits. On June 9th, 2011 the *Indian Express* reported:

Declaring that "science is politics in climate change; climate science is politics," Union Environment Minister Jairam Ramesh on Wednesday urged Indian scientists to undertake more studies and publish them vigorously to prevent India and other developing countries from being "led by our noses by Western (climate) scientists who have less of a scientific agenda and more of a political agenda."[16]

This was also framed in the context of priorities—economic growth is more important. Ramesh said:

But we can't take on absolute emission cuts like the Europeans—firstly because we didn't cause the problem of global warming, and secondly because we have a huge developmental backlog. For example, we have 300-400 million people to whom we have to extend electricity.[17]

Many necessary changes have occurred to our awareness of the environment. We require many more, but extremists demand a complete and unsustainable restructuring of world economies in the guise of environmentalism. Society is beginning to realize there are reasonable limits good for the earth well short of human eradication.

People are tired of the eco-bullying and unnecessary disruptions to their economies and their lives. The extremists continue to bully but fail because their threats of disaster fail. The methods they use to bully are exposed. The challenge for many,

---

[16] http://www.indianexpress.com/news/Climate-science-mired-in-politics--Jairam/801199/
[17]
http://content.time.com/time/world/article/0,8599,2025113,00.html

particularly sensible leaders, becomes preserving the benefits and values in the exploited paradigm.

These challenges are facing climate change and environmentalism. The Pew poll (Table 1) in the U.S. has for a few years now shown global warming and environmentalism very low on the list of public concern. The table shows concerns in 2010. Notice the biggest change was in Environment and Global warming. It will get worse when the public discover how they were misled on these issues. The challenge is not to throw the baby out with the bathwater.

Top Policy Priorities for 2012

% considering each as a "top priority" for the president and Congress this year	Five years ago Jan 2007 %	One year ago Jan 2011 %	Today Jan 2012 %	Five year chg
Economy	68	87	86	+18
Jobs	57	84	82	+25
Terrorism	80	73	69	-11
Budget deficit	53	64	69	+16
Social Security	64	66	68	
Education	69	66	65	
Medicare	63	61	61	
Tax fairness	--	--	61	
Health care costs	68	61	60	-8
Energy	57	50	52	
Help poor and needy	55	52	52	
Crime	62	44	48	-14
Moral breakdown	47	43	44	
Environment	57	40	43	-14
Lobbyist influence	35	37	40	
Illegal immigration	55	46	39	-16
Strengthening military	46	43	39	-7
Global trade	34	34	38	
Transportation	--	33	30	
Lower military spending	--	--	29	
Campaign finance	24*	--	28	
Global warming	38	26	25	-13

PEW RESEARCH CENTER Jan. 11-16, 2012.
* Campaign finance reform trend from Jan. 2004.

Table 1

Foolishly we've developed global energy and economic policies based on incorrect science promulgated by extremists, though it serves to define the limits. As usual, cautionary red flags were everywhere, but a combination of deliberate misdirection by some politicians and scientists—amplified by the mainstream media—fuelled the insanity. Questioners are bullied into silence with accusations that they don't care about the planet, the future or the children. The extremists usurped the moral high ground.

In the moral vacuum created by the defeat of religion by science, many sought a new belief system. Environmentalism fits the bill. It hearkened back to the worship of nature, known as animism, of non-Christian societies. Ironically, in becoming the new religion, environmentalism became dogmatic like all religions.

So, in the western world we moved from the dogmatism of Christianity to the dogmatism of science and then to the dogmatism of environmentalism. It is unsurprising that Sir John Houghton, first co-chair of the Intergovernmental Panel on Climate Change (IPCC) and lead editor of the first three IPCC Reports, confronted the dilemmas in an article for *The Global Conversation*:

> First let me write a few words about God and science. A few prominent scientists are telling us that God does not exist and science is the only story there is to tell. To argue like that, however, is to demonstrate a fundamental misunderstanding of what science is about. At the basis of all scientific work are the 'laws' of nature—for instance, the laws of gravity, thermodynamics and electromagnetism, and the puzzling concepts and mathematics of quantum mechanics. Where do these laws come from? Scientists don't invent them; they are there to be discovered. With God as Creator, they are God's laws and the science we do is God's

science.

> *The Earth is the Lord's and everything in it (Psalm 24), and Jesus is the agent and redeemer of all creation (John 1:2; Colossians 1:16-20; Ephesians 1:16). As we, made in God's image, explore the structure of the universe that God has made with all its fascination, wonder and potential, we are engaging in a God given activity. Many of the founders of modern science three or four hundred years ago were Christians pursuing science for the glory of God. I and many other scientists today are privileged to follow in their footsteps.*
>
> *A special responsibility that God has given to humans, created in His image, is to look after and care for creation (Genesis 2:15). Today the impacts of unsustainable use of resources, rapidly increasing human population and the threat of climate change almost certainly add up to the largest and most urgent challenge the world has ever had to face—all of us are involved in the challenge, whether as scientists, policy makers, Christians or whoever we are.*[18]

Natural paradigm shifts are not smooth. There is always potential for revolution. The shift to environmentalism was hijacked for a political agenda.

---

[18] Sir John Houghton, *The Science of Global Climate Change. Facing the Issues. What are the Issues?* Cape Town 2010 Advance Paper

MAURICE STRONG

CHAPTER THREE

Population—Overpopulation

PLATO AND ARISTOTLE discussed the ideal size for a city-state in the 3rd and 4th centuries BC, but Confucius preceded them in the 5th century with warnings about excessive growth. Predictions of population outgrowing resources received a boost by the speculations of Thomas Malthus in 1798 with his work *An Essay on the Principle of Population*. It was Malthus' notion of too-many-humans-for-too-little-food that galvanized the modern debate. The Club of Rome expanded this to the idea that an industrialized population would outgrow all resources.

The most recent flurry of alarmism over population growth was key to the ideas of the Club of Rome, which received momentum through Paul Ehrlich's even more egregious and inaccurate book *The Population Bomb*.[19] The error of their predictions didn't stop extremists seeing the need for total control. Ehrlich published with John Holdren and they advanced the idea of limiting or reducing population.

As environmental extremists go, Holdren and Ehrlich are not so extreme: they think only a limited number of people should exist. The problems are: what is the right number and who should

---

[19] *The Population Bomb*, Dr. Paul R. Ehrlich, Sierra Club Press, San Francisco, 1969

27

determine it? Shall they decide there are too many of certain races or abilities? They, along with Anne Ehrlich, co-authored a 1977 book titled *Ecoscience: Population, Resources, and Environment*. One quote illustrates the dangers.

> *Indeed, it has been concluded that compulsory population control laws, you even including laws requiring compulsory abortion, could be sustained under the existing Constitution if the population crisis became sufficiently severe to endanger the society.*[20]

But who determines that the crisis endangers society?

Limits To Growth was a 1972 report produced for The Club of Rome. It used grossly simplistic linear models to claim the world was on the verge of collapse. This pattern of using models to create a predetermined outcome anticipated the climate models later created by the IPCC. The central theme was a continuation of the idea about population capacity of the world that has gone on throughout written history.

It followed and built on that which already claimed the world was overpopulated and doomed. We must have population control at home, hopefully through a system of incentives and penalties, but by compulsion if voluntary methods fail. Some categorize Ehrlich's predictions among the most ridiculous on record. Consider just three, any one of which should raise flags about the author's credibility.

1. *The battle to feed humanity is over. In the 1970s, the world will undergo famines. Hundreds of millions of people are going to starve to death in spite of any crash programs embarked upon now.*

---

[20] *Ecoscience: Population, Resources, Environment*, Paul R. Ehrlich, John P. Holdren, Anne H. Erhlich, W.H. Freeman & Co., 1978.

2. *Four billion people—including 65 million Americans—would perish from famine in the 1980s.*

3. *In ten years [i.e., 1980] all important animal life in the sea will be extinct. Large areas of coastline will have to be evacuated because of the stench of dead fish.*

It's a measure of the almost religious fervor of the environmental paradigm that such failed predictions, which would result in total rejection of the hypothesis in any other branch of science, continue. Despite the ludicrous nature of these claims, environmentalists and people who saw the political potential of the fears these claims engendered promoted them. John Holdren co-authored a 1969 article with Ehrlich that claimed:

> *...if the population control measures are not initiated immediately, and effectively, all the technology man can bring to bear will not fend off the misery to come.*[21]

Holdren, as White House Science Czar, was in a position to make the prophecy reality. He was a fully paid-up member of The Limits to Growth club. For example, in his 1971 Sierra Club book, *Energy: A Crisis in Power*, he declared:

> *...it is fair to conclude that under almost any assumptions, the supplies of crude petroleum and natural gas are severely limited. The bulk of energy likely to flow from these sources may have been tapped within the*

---

[21] *Population and Panaceas A Technological Perspective,* Paul R. Erlich and John P. Holdren, *BioScience,* Vol. 19, No. 12 (Dec., 1969), pp. 1065-1071, http://www.jstor.org/stable/1294858

lifetime of many of the present population.[22]

Because of this view, Holdren joined with Ehrlich and John Harte in a bet[23] with economist Julian Simon that five metals would increase in price over ten years because of increasing scarcity. Holdren carried out the task of selecting the metals and the time period in which it would occur. Simon easily won the bet.

What is the reality of overpopulation? Is the claim the ultimate application of the acronym Not In My Back Yard (NIMBY)? It is certainly at the heart of the anti-humanity view of extreme environmentalists—get rid of all the people and the planet would be a good place to live. A few people claim overpopulation is the primary cause of global warming and climate change, but this only underscores a lack of knowledge of the issues. It is also the ultimate form of political control[24] as history illustrates.

Global warming due to overpopulation accepts the assumption that human addition of CO_2 to the atmosphere is the cause and that adding more people will produce more CO_2. Not only is this claim wrong, but also CO_2 in total is not the cause. How many times does it have to be said that temperature increases before CO_2 in every single record? For some, it becomes a form of the precautionary principle: the idea that, even if humans are not causing warming, shouldn't we limit population anyway?

The world is not overpopulated. The U.S. Census Bureau

---

[22] Energy: A Crisis in Power, John P. Holdren and Philip Herrera, Sierra Club Press, 1972.

[23] http://www.forbes.com/2009/02/03/holdren-obama-science-opinions-contributors_0203_ronald_bailey.html

[24] http://thecriticalthinker.wordpress.com/2008/11/01/ted-turner-overpopulation-causes-global-warming/

provides a running estimate of population.[25] Note that this is an estimate, because no accurate census exists for any country. This includes the U.S. that spends more money and effort[26] than any other nation.

Most of the world is essentially unoccupied; most of our population is concentrated in coastal flood plains and deltas. Consider that Canada is the second largest country in the world and has approximately 33.6 million residents.[27] Compare this with California with a 2008[28] population of an estimated 36.8 million.

Calculations often appear that put the entire world population on a certain islands or in a specific region. Do the math with Texas at 7,438,152,268,800 square feet divided by 6,774,436,692 (2009 estimate) for a density of 1098 square feet per person. Fitting all the people in an area is different from their being able to live there. Population geographers distinguish between ecumene, the inhabited area and non-ecumene, the uninhabited areas. The problem is the definition of what are habitable changes all the time as areas previously uninhabitable become habitable. Similarly, the area of the earth that is habitable has changed because of technology, communications and food production capacity. Three great turning points in human history identified by anthropologists are all related to weather and climate control. Use of fire and clothing allowed survival in colder regions while irrigation offset droughts and allowed settlement in arid regions.

Regardless of the real world situation, the issue was always the carrying capacity of the land—an issue often twisted in service

---

[25] http://www.census.gov/main/www/popclock.html

[26] http://usatoday30.usatoday.com/news/nation/census/2009-07-26-overcount_N.htm

[27] Source: http://www.statcan.gc.ca/start-debut-eng.html

[28] Source:
http://factfinder2.census.gov/faces/nav/jsf/pages/index.xhtml

of political power by a left wing ideology. Ehrlich blatantly said in his book *The Population Bomb* that humans occupy only three percent of the earth's land surface. So his overpopulation concern is invalid from a carrying-capacity perspective.

It was the activities of a portion of the population that offended his political views. In an April 6th, 1990, Associated Press quote he said:

> *Actually, the problem in the world is there are too many rich people.*[29]

Compare his comment, quoted in Dixy Lee Ray's book *Trashing the Planet*:

> *We've already had too much economic growth in the United States. Economic growth in rich countries like ours is the disease not the cure.*[30]

with Maurice Strong's comment:

> *Isn't the only hope for the planet that the industrialized nations collapse? Isn't it our responsibility to bring that about?*[31]

The difference is Strong acted on his belief, as we will see.

---

[29] *Population Expert Faults Wealthy,* Sarasota Herald-Tribune, April 6, 1990,
http://news.google.com/newspapers?id=TjAcAAAAIBAJ&sjid=KHoE AAAAIBAJ&pg=6924,333924&hl=en

[30] Trashing the Planet: How Science Can Help Us Deal With Acid Rain, Depletion of the Ozone, and Nuclear Waste (Among Other Things), Dixy Lee Ray, Perennial Press, 1992.

[31] Maurice Strong, Opening speech, Rio Earth Summit, 1992

Paul Ehrlich's credibility suffered, but the idea of overpopulation had gained political momentum. By 1994, his predictions were already completely wrong, but this did not stop the United Nations, with the enthusiastic support and participation of Al Gore, from holding a population conference in Cairo, Egypt.[32] What better place to hold the conference with the teeming, hungry millions outside the conference hall? They ignored the fact that the Netherlands—with among the highest population density—also had among the highest standard of living.

There was another connection with Maurice Strong because the conference grew out of the 1992 Rio conference he organized. As the UN information from the Cairo population Conference[33] notes:

> *The Rio Declaration on Environment and Development and Agenda 21, adopted by the international community at the United Nations Conference on Environment and Development, call for patterns of development that reflect the new understanding of these and other intersectoral linkages.*[34]

This is bureaucratic justification for tying population to other perceived evils like global warming and to achieve sustainable development. The latter is the perfect political phrase that means everything to everyone and nothing to anyone. They did it with

---

[32]Source:http://www.ibiblio.org/pub/archives/whitehouse-papers/1994/Sep/1994-09-13-VP-Gore-on-Close-of-Population-and-Development-Conf

[33] The United Nations International Conference on Population and Development (ICPD) held from 5-13 September 1994 in Cairo, Egypt. http://www.iisd.ca/cairo.html

[34] Source:http://www.iisd.ca/Cairo/program/p03001.html

this statement:

> *Explicitly integrating population into economic and development strategies will both speed up the pace of sustainable development and poverty alleviation and contribute to the achievement of population objectives and an improved quality of life of the population.* [35]

Al Gore, head of the U.S. delegation put it more succinctly in his statement at the Cairo conference when he wrote:

> *No single solution will be sufficient by itself to produce the patterns of change so badly needed. But together, over a sufficient length of time, a broad-based strategy will help us achieve a stabilized population and thereby improve the quality of life for our children. The Program of Action just adopted in Cairo offers us a plan that will work and that has the full support of the United States.* [36]

In 1976, Strong told the Canadian magazine Maclean's:

> *I am a socialist in ideology, a capitalist in methodology.*

Therefore we shouldn't be surprised he is making a great deal of money from exploiting the false doctrine of human induced climate change. The fact that the ideology precludes the methodology doesn't bother a master manipulator like Strong. Ronald Bailey provides the following quote in his September 1st, 1997 *National Review* article about Strong:

---

[35] Source: http://www.unfpa.org/pds/sustainability.htm

[36] U.S. Department of State Dispatch Supplement Volume 5, Number, September 8th, 1994, UN International Conference on Population and Development,

He's dangerous because he's a much smarter and shrewder man [than many in the UN system]...

Charles Lichenstein, deputy ambassador to the UN under President Reagan said of Strong.

I think he is a very dangerous ideologue, way over to the Left.

Al Gore is different in that his motive was, initially, purely political. The money came later. Gore used—or rather misused—the misleading information of the IPCC, which is why he shared the Nobel Prize with them. However, as his political ambitions receded, he also began making a great deal of money through his involvement with carbon credit trading. Scientists of the IPCC may be involved in carbon trading, but they also benefit through a high profile, making access to funding and promotion easier. Relationships between Gore, Strong and carbon credit trading are well documented.[37]

Failures Fail To Destroy Credibility

Despite this, the credibility of these people remained unsullied and essentially unchallenged in the same way that the credibility of the CRU/IPCC does today. A significant reason was because they both used computer models that mystify and beguile most people, as Pierre Gallois quipped:

If you put tomfoolery into a computer, nothing comes out of it but tomfoolery. But this tomfoolery, having passed through a very expensive machine, is somehow ennobled and no-one dares criticize it.

---

[37] http://capitalresearch.org/pubs/pdf/v1185475433.pdf

However, beyond not understanding the mathematics, the public believed that the models produced accurate predictions of the future. In fact, the *Limits to Growth World3* model[38] was not intended to be predictive. *"In this first simple world model, we are interested only in the broad behavior modes of the population-capital system."* The *"population-capital"* phrase is telling because it links the catastrophic population predictions of Ehrlich with the economic and political system. It was a barely disguised attack upon the capitalist system. It set a pattern because IPCC models do the same thing. Most think, incorrectly, they are only about global temperature. The final Reports produced by Working Group II[39] evolve from the temperature predictions of Working Group I.

> *After confirming in the first volume on "The Physical Science Basis" that climate change is occurring now, mostly as a result of human activities, this volume illustrates the impacts of global warming already under way and the potential for adaptation to reduce the vulnerability to, and risks of climate change.*

They blend them into the influential Summary for Policy Makers (SPM) that claims population and economic growth, measured by CO_2, have caused temperature increase and the solutions are to stop economic growth and reduce the population.

Models of both groups use population projections and both assume business as usual. Both are simplistic linear models based on virtually no data, with wholly inadequate understanding of the mechanisms and inabilities to accommodate feedback. Every model prediction or projection of both groups was remarkably wrong, regardless of attempts at improvement. This is true for

---

[38]Source: http://dieoff.org/page25.htm
[39]Source:http://www.ipcc.ch/publications_and_data/ar4/wg2/en/frontmattersforeword.html

weather, climate, population and economic models.

A similar record of failed predictions would doom any other area of research and policy. Money and politics combined with the use of computer models to bestow an entirely unjustified credibility.

It's time to expose their failures to the public before their work does too much more damage.

CHAPTER FOUR

Transition from The Club Of Rome to the United Nations

IN THE POLITICAL climate engendered by environmentalism and its exploitation, some demand a new world order and they believe this can be achieved by shutting down the industrialized nations. It was a major theme of The Club of Rome and driven by studies like *Limits to Growth* and Paul Ehrlich's book *The Population Bomb*. Maurice Strong was a senior member of the Club and speculated in 1990:

> *What if a small group of these world leaders were to conclude the principal risk to the earth comes from the actions of the rich countries?...In order to save the planet, the group decides: Isn't the only hope for the planet that the industrialized civilizations collapse? Isn't it our responsibility to bring this about?*[40]

Strong's speculation became a challenge taken up by the Club that

---

[40] http://www.answers.com/topic/maurice-strong

required translation to a workable plan.

Consider the challenge of collapsing an industrialized civilization. This is where carbon dioxide (CO_2) becomes the focus. The civilization the Club opposed comprised nations built on and driven by the energy provided by fossil fuels. It's reasonable to compare these nations to a car, the very symbol of all they detest. You can stop a car engine by cutting off the fuel supply, but that would be extremely difficult and elicit quick anger in a country, as anger when fuel prices jump demonstrate. However, you can also stop a car engine by blocking the exhaust. Transfer that idea to nations and show that CO_2, the byproduct of combustion of fossil fuels, was causing runaway, catastrophic, global warming to achieve the goal. What nastier image than the belching car exhaust or the even more dramatic chimneys of industry?

On page 75 of their Report for the Club, The First Global Revolution,[41] Alexander King and Bertrand Schneider wrote:

> *In searching for a common enemy against whom we can unite, we came up with the idea that pollution, the threat of global warming, water shortages famine and the like, would fit the bill. In their totality and their interaction these phenomena do constitute a common threat which must be confronted by everyone together. But in designating these dangers as the enemy, we fall into the trap, which we have already warned readers about, namely mistaking symptoms for causes. All these dangers are caused by human intervention in natural processes, and it is only through changed attitudes and behaviour that they can be overcome. The real enemy then is humanity itself.*

---

[41] http://www.scribd.com/doc/13160503/The-First-Global-Revolution-Club-of-Romes-1991-Report

An ambitious plan, but how is it implemented? Nobody was more skilled at manipulating such a plan than Maurice Strong. When asked by Elaine Dewar why he didn't enter politics to implement his plan to get rid of the industrialized civilizations, he essentially said you couldn't do anything as a politician. However, he knew you needed a political vehicle to achieve the goal. It is almost impossible to convince all governments separately as Kyoto and current climate negotiations prove. His experience told him the United Nations (UN) was his vehicle.

Elaine Dewar wrote about Strong in her book *Cloak of Green* and concluded that he liked the UN because:

> *He could raise his own money from whomever he liked, appoint anyone he wanted, control the agenda.*[42]

As Dewar later wrote:

> *Strong was using the U.N. as a platform to sell a global environment crisis and the Global Governance Agenda.*

Strong is most responsible for setting up the bureaucratic structure necessary to control the political and science agendas. Neil Hrab wrote in 2001[43] he achieved this by:

> *Mainly using his prodigious skills as a networker. Over a lifetime of mixing private sector career success with stints in government and international groups...*

This began with the 1977 United Nations Conference on the

---

[42] *Cloak of Green: The Links between Key Environmental Groups, Government and Big Business,* Elaine Dewar, Lorimer Press, 1995.
[43] http://www.enterstageright.com/archive/articles/1201/1201strong.txt

Human Environment Stockholm Conference. Berry says the Conference was simply a success. As Hrab notes:

> *The three specific goals set out by the Secretary General of the Conference, Maurice F. Strong, at its first plenary session—a Declaration on the human environment, an Action Plan, and an organizational structure supported by a World Environment Fund—were all adopted by the Conference.*[44]

Hrab also noted:

> *What's truly alarming about Maurice Strong is his actual record. Strong's persistent calls for an international mobilization to combat environmental calamities, even when they are exaggerated (population growth) or scientifically unproven (global warming), have set the world's environmental agenda.*

It confirms the accuracy of Elaine Dewar's label *The Global Governance Agenda*.

Other events were providing the fertile social and political ground needed to further the goals. Anything that would suggest human activities and particularly industry were causing environmental problems became a focus. As discussed, The Club of Rome, created in April of 1968, published *Limits to Growth* and *The Population Bomb* to academically legitimize fear. A report Strong commissioned for the first UNEP conference and prepared by Barbara Ward and Rene Dubos titled, *Only One Earth: The Care and Maintenance of a Small Planet* essentially became the first state of the environment report. It reinforced the shrinking planet perspective provided by the Apollo 8.

---

[44] Bulletin of the Atomic Scientists, September, 1972. p.17

Wonderful political catch phrases appeared such as Dubos' *Think globally, act locally* or the Brundtland Commission's *Sustainable development,* which spread quickly and became part of the vernacular. The latter phrase was a typical vague political statement meaning everything to everyone, but nothing to anyone. Development came to mean constant growth and in that context is unsustainable. It was an oxymoron prefacing a series of such contradictions emerging as politics and emotion overtook science and logic. It also gave extreme environmentalists the moral high ground, a position from which they bully society and suppress scientists who dare to question.

These groups received a world platform and ascendancy through Consultative Status at the 1992 conference Strong organized and chaired in Rio de Janeiro. The idea of Consultative Status evolved, along with the concept of Non-Governmental Organizations (NGO) from original ideas incorporated in the UN Charter. The conference, called the Earth Summit, followed the practice of essentially excluding large segments of society including industry and business. Subsequently, they received token status by the establishment of the World Business Council for Sustainable Development (WBSCD), but who has heard of them? They did add a critical piece at the Conference for the Climate Change Convention out of which the Kyoto Accord emerged. It furthered Strong's agenda of controlling climate science through politics.

Now the structures were in place to control the science and further the political agenda; policies could evolve, but because they were based on incorrect science, would have devastating consequences. Now, the challenge was to perpetuate the misinformation and divert scientists who despite personal attacks, denial of funding and exclusion from national and world level conferences continued to pursue the scientific method.

The challenge was twofold. Advance the political agenda and produce the scientific evidence to provide legitimacy.

Chapter Four

Organization of, and appointment as first Secretary General of, the United Nations Environmental Program established in 1972 provided the political platform. Out of that agency and in conjunction with the World Meteorological Organization (WMO), they created the IPCC to provide and advance the scientific evidence. This is the group touted as the consensus on climate change research. It is anything but, and has been a political agency from its inception. By publishing periodic reports, it convinced the public that humans—via their CO_2 emissions—cause harmful climate change. They became the official source on climate.

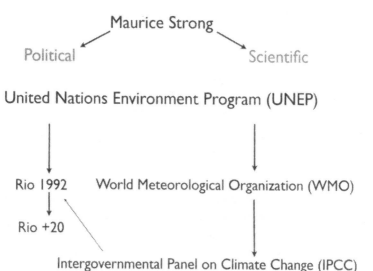

Maurice Strong

Political — Scientific

United Nations Environment Program (UNEP)

Rio 1992 World Meteorological Organization (WMO)

Rio +20

Intergovernmental Panel on Climate Change (IPCC)

Strong's first significant move was establishing the United Nations Environmental Program (UNEP), which emerged from the 1972 Stockholm conference. From there they cut their teeth on the ozone issue culminating in the Montreal Protocol in 1987. Examining the parallels between the ozone and CO_2 issues reveal a great deal. For example, the focus was a product of human activities, chlorofluorocarbons (CFCs) that caused damage to the atmosphere. Science manufactured proof of CFCs destroying ozone, when it didn't happen in the Ozone Layer. Later the Montreal Protocol claimed as successful was exposed as a charade because ozone variation was due to natural causes. The ultraviolet portion of sunlight creates ozone, but because they assumed sunlight was constant they eliminated the actual cause of ozone variation.

Once established, UNEP set out the political objective at the Rio Earth Summit in 1992. Titled Agenda 21, an abbreviation for a plan for the 21st century it set out in its Principles a template for achieving global governance and transfer of wealth. Among them was Principle 15 that effectively waives the need for science:

> *In order to protect the environment, the precautionary approach shall be widely applied by States according to their capabilities. Where there are threats of serious or irreversible damage, lack of full scientific certainty shall not be used as a reason for postponing cost-effective measures to prevent environmental degradation.*[45]

This principle incorporates all the politics of the Agenda. It is essentially the Precautionary Principle, a favorite tool of environmentalists to bypass the need for facts as the basis of decision-making. It says: if you're not sure, act anyway. It's incorporated in Principle 15 by the phrase "lack of full certainty." It also builds on the aim of wealth distribution because it limits action to those who have the capability. The wealthy have to pay for their sins.

UNEP worked with the World Meteorological Organization (WMO) to form the IPCC. The objective was to produce the

---

[45] http://www.gdrc.org/u-gov/precaution-7.html

science that human CO_2 was causing global warming. They began with the direction to use the definition of climate change approved by the United Nations Framework Convention on Climate Change (UNFCCC).

Article 1 of the UNFCCC, a treaty formalized at the "Earth Summit" in Rio in 1992, defined Climate Change as:

> *a change of climate which is attributed directly or indirectly to human activity that alters the composition of the global atmosphere and which is in addition to natural climate variability observed over considerable time periods.*[46]

This makes the human impact the primary purpose of the research. The problem is you cannot determine human contribution unless you know the amount and cause of natural climate change. Hubert Lamb, who many consider the father of modern climate studies, was the founder and first Director of the Climatic Research Unit (CRU) said in his autobiography *Through All the Changing Scenes of Life* that he created the Climatic Research Unit (CRU) because:

> *...it was clear that the first and greatest need was to establish the facts of the past record of the natural climate in times before any side effects of human activities could well be important.*[47]

Properly, a scientific definition would put natural climate variability first, but at no point does the UN mandate require an

---

[46]http://unfccc.int/essential_background/convention/background/items/2536.php

[47] *Through all the Changing Scenes of Life—A Meteorologist's Tale,* Hubert Lamb, Taverner Publications, 1997.p. 203.

advance of all climate science. The definition used by UNFCCC predetermined how the research and results would be political and produce pre-determined results. It made discovering a clear *human signal* mandatory, but meaningless. It also thwarted the scientific method.

They defined climate change in 1992 after formalizing the IPCC in 1988. Strong had two objectives: create the science needed to prove human CO_2 was the problem and then convince the public if they didn't act the outcome would be catastrophic. He needed control of selecting participants, especially Lead Authors and did it through the World Meteorological Organization. It is no coincidence that the Chair of the 1985 meeting was Gordon McBean, Assistant Deputy Minister of Environment Canada. As Richard Lindzen explained:

> *IPCC's emphasis, however, isn't on getting qualified scientists, but on getting representatives from over 100 countries, said Lindzen. The truth is only a handful of countries do quality climate research. Most of the so-called experts served merely to pad the numbers.*[48]

Using Weather Departments gave the bureaucrats ascendancy over politicians because to challenge put them in contradiction with their own experts. This controlled the flow of information in each country. Again Lindzen from his direct involvement with the IPCC wrote:

> *It is no small matter that routine weather service functionaries from New Zealand to Tanzania are referred to as 'the world's leading climate scientists.' It should come as no surprise that they will be determinedly*

---

[48] http://news.heartland.org/newspaper-article/2001/06/01/ipcc-report-criticized-one-its-lead-authors

supportive of the process.[49]

A political bias made a few of them especially supportive.

More important than the political control of climate science through bureaucracies was control of climate research funding in each country. Joanne Nova[50] documented the commitment of $79 billion between 1989 and 2009 by the U.S. government. However, the amount is not the critical scientific issue. They directed virtually all the funding to research and researchers proving the AGW hypothesis. As Nova notes:

> *Thousands of scientists have been funded to find a connection between human carbon emissions and the climate. Hardly any have been funded to find the opposite. Throw 30 billion dollars at one question and how could bright, dedicated people not find 800 pages worth of connections, links, predictions, projections and scenarios? (What's amazing is what they haven't found: empirical evidence.)*[51]

This further thwarted the scientific method that tries to disprove the hypothesis. It resulted in the disproportionate number of papers on one side of the debate, which AGW proponents used as another consensus argument.

Other parts of their mandate illustrate the political nature of the entire exercise. Its principles require the IPCC:

> *...shall concentrate its activities on the tasks allotted to*

---

[49] Ibid.

50

http://scienceandpublicpolicy.org/images/stories/papers/originals/climate_money.pdf

[51] Ibid.

*it by the relevant WMO Executive Council and UNEP
Governing Council resolutions and decisions as well as on
actions in support of the UN Framework Convention on
Climate Change.*[52]

The role is also to:

*...assess on a comprehensive, objective, open and
transparent basis the scientific, technical and socio-
economic information relevant to understanding the
scientific basis of risk of human-induced climate change,
its potential impacts and options for adaptation and
mitigation. IPCC reports should be neutral with respect to
policy...*[53]

They designed instructions to consider the socio-economic
impacts in the structure, which provided further bias. It takes the
research further away from climate science. This became apparent
when they introduced the *consensus* argument. Consensus is
neither a scientific fact nor important in science, but it is very
important in politics.

There are 2,500 members in the IPCC divided between 600
in Working Group I (WGI) who produce a Science Report and
1,900 in Working Groups II and III (WG II and III) all studying
impacts. Of the 600 in WGI, only 308 worked on the science part
of the report and only five reviewed all 11 chapters. The 1,900 in
the other groups accepted without question the findings of WGI
and assumed warming due to humans was a certainty.

In a circular argument, typical of so much climate politics,

---

[52] From: Principles Governing IPCC work, approved at the 14th
Session, Vienna 1-3 October 1998 and amended at the 21st Session,
Vienna 6-7 November, 2003.
[53] Ibid.

they present the work of the 1900 as *proof* of human caused global warming. Working Group II speculated on the negative impacts of a warming world. They didn't consider benefits as a normal cost-benefit analysis requires. Worse, the speculations become facts in the sensationalist media. This pattern of only considering costs was normal and most egregious in the Stern Report[54] commissioned by the British government in 2006. The Report received extensive coverage in the mainstream media and presented almost as fact when it was pure speculation based on the IPCC SPM.

The cynicism of the phrase "IPCC reports should be neutral with respect to policy..."[55] underscores the deception in the entire process. They cannot be neutral because the Science is deliberately not neutral. They then corrupt and distort neutrality by producing the Summary for Policymakers (SPM), the most important part of the IPCC work, which is always dramatically different than the Science Report. It is hard to believe that any intelligent person participating in the IPCC process can read the Science Report of Working Group I and the SPM and not realize they are completely different assessments.

Part of the procedure to ensure people, especially the media, would read the SPM and not the Science Report was to ensure its release months before the Science Report. It is very likely that few people ever read the Science Report. It is more logical to produce the Science Report first and then the SPM. There is only one explanation for producing it first. The final product achieved the result of deception in full daylight because as David Wojick, IPCC expert reviewer, explained:

---

[54] http://www.guardian.co.uk/politics/2006/oct/30/economy.uk

[55] http://www.ipcc.ch/organization/organization_history.shtml#.UcXIxPmsh8E

Glaring omissions are only glaring to experts, so the "policymakers"—including the press and the public—who read the SPM will not realize they are being told only one side of a story. But the scientists who drafted the SPM know the truth, as revealed by the sometimes artful way they conceal it.

...

What is systematically omitted from the SPM are precisely the uncertainties and positive counter evidence that might negate the human interference theory. Instead of assessing these objections, the Summary confidently asserts just those findings that support its case. In short, this is advocacy, not assessment.[56]

It is worse because the Summary does not identify the degree of difference between the Science Report and SPM. Perhaps the most egregious example is the chapter on Climate Models. Programmed to produce a result they are another of the classic climate science circular arguments. They program them so a CO_2 increase causes a temperature increase. Then when the results show an increase in temperatures, they claim it proves the assumption.

It was no accident that whoever leaked the emails from the Climatic Research Unit (CRU) also leaked computer codes. All who have examined them say they are a mess and as one person commented:

...the emails are only about 5 percent of the total. What does examining the other 95 percent tell us? Here's the short answer: it tells us that something went very wrong

---

[56] http://www.john-daly.com/guests/un_ipcc.htm

in the data management at the Climatic Research Unit.[57]

Beyond the management, there are serious problems with the limited amount of data and the number of variables omitted.

---

[57] http://pjmedia.com/blog/climategate-computer-codes-are-the-real-story/

JAMES HANSEN

CHAPTER FIVE

The Search for a Human Signal and Political Machinations Designed to Prove Human CO_2 was Causing Global Warming

We've got to ride the global warming issue. Even if the theory of global warming is wrong, we will be doing the right thing...
—Senator Tim Wirth, 1993[58]

JAMES HANSEN, DIRECTOR of NASA-GISS became central to the deception about global warming when he took it from behind the scenes of the UN bureaucratic machinations into the political, public and mainstream media spotlight. His June 1988 appearance before a U.S. Senate Committee, at which he claimed he was certain that human CO_2 caused global warming, was an orchestrated event. The maestros of this deception were proud of what they did.

---

[58] As related in *Science under Siege: How the Environmental Misinformation Campaign Is Affecting Our Lives*, Michael Fumento, William Morrow, June 1996.

Directors of the 1988 drama were Al Gore and Timothy Wirth. In a 2007 PBS Frontline documentary Wirth said:

> *We knew there was this scientist at NASA, you know, who had really identified the human impact before anybody else had done so and was very certain about it. So we called him up and asked him if he would testify.*[59]

Then they took actions that illustrate why people distrust and despise most politicians. It also shows they would go to virtually any length to achieve their goal.

> *We called the Weather Bureau and found out what historically was the hottest day of the summer. Well, it was June 6th or June 9th or whatever it was. So we scheduled the hearing that day, and bingo, it was the hottest day on record in Washington, or close to it.*[60]

The interviewer asked Wirth:

> *Did you also alter the temperature in the hearing room that day?*[61]

Wirth replied:

> *What we did is that we went in the night before and opened all the windows, I will admit, right, so that the*

---

[59]

http://www.pbs.org/wgbh/pages/frontline/hotpolitics/interviews/wirth.html

[60] Ibid.

[61] Ibid.

air conditioning wasn't working inside the room.[62]

Since that first deliberate deception Hansen has tried to confirm his claim that:

> *...the global warming is now large enough that we can ascribe, with a high degree of confidence, a cause-and-effect relationship to the greenhouse effect.*[63]

Hansen's statement before the Gore committee[64] stating that human CO_2 was causing warming, is unsupportable, especially with the certainty he claimed, is one no honest scientist would make, but is what the politicians wanted. He reportedly admitted to Stephen Schneider (now deceased) he should not have made it,[65] but he did not go public and admit his error and certainly didn't go back to the committee—instead, he became a science adviser to Al Gore.

I've discussed how Maurice Strong hijacked environmentalism and climate science to achieve the political goal

---

[62] Ibid.

[63] http://image.guardian.co.uk/sys-files/Environment/documents/2008/06/23/ClimateChangeHearing1988.pdf

[64] United States Senate Committee on Energy and Natural Resources, June 23rd, 1988

[65] Some scientists, such as James Hansen and myself, have led efforts to estimate what volcanic or solar forcings may have contributed to 20th century temperature trends. Indeed, such estimates improve the match of our computer model simulations to observed twentieth-century temperature trends. However, all of us have admitted, without more quantitatively reliable information these exercises can do nothing more than sketch out plausible rather than definitive results.
—Stephen H. Schneider, *Global Warming, Are We Entering the Greenhouse Century?*, Lutterworth Press, 1989, p. 297

of causing the demise of industrialized nations. Strong established the political vehicle, the UNEP, and the scientific vehicle, the IPCC, for his purpose. He brought them together at the Earth Summit in Rio in 1992. The fruits of his efforts and the policies they engendered are now emerging and are hurting the poor and middle-income people of all countries, with rising food and energy costs. They're hurting the people they claimed to help, but more on that later.

A sign that the objectives were about politics not science was signaled early by Sir John Houghton, first co-chair of the IPCC and lead editor of its first three Reports, when he said:

> *If we want a good environmental policy in the future, we'll have to have a disaster. It's like safety on public transport. The only way humans will act is if there's been an accident.*[66]

The IPCC has done this with ruthless efficiency while pretending what they are doing is science not politics. Houghton's[67] excessive statement "...the impacts of global warming are like a weapon of mass destruction..." precedes the claim that it kills more people than terrorism.

The trouble is: more people die of cold weather each year than warm. Also, notice the word *impact* because that, not science, dominates the work of the IPCC. Three quarters of the people involved in the IPCC (1,900 of 2,500) study what *might happen*, not will happen. So they established the entire process to achieve the goal of announcing (potential) disasters.

Bert Bolin was Houghton's fellow co-chair. Bolin had a

---

[66] *Me and my God* column, *Sunday Telegraph*, September 10th, 1995

[67]

http://www.guardian.co.uk/politics/2003/jul/28/environment.greenpolitics

history of involvement in the politics of the environment. Both he and Houghton signed the 1992 warning to humanity essentially blaming the developed nations. It was a typical Club of Rome approach with no clear measures or evidence, simply a list of possible disasters if we didn't do things their way.

Science works by creating theories based on assumptions, which other scientists, performing as skeptics, test. The structure and mandate of the IPCC was in direct contradiction to this scientific method. They set out to prove the theory rather than disprove it. Maurice Strong and his UN committees made sure the focus was on human-caused change and CO_2 as the particular culprit. They'd already biased the research by using a very narrow definition of climate change as discussed earlier. Properly, a scientific definition would put natural climate variability first, but at no point does the UN mandate require an advance of climate science. The definition used by UNFCCC predetermined how the research and results would be political and pre-determined. It made discovering a clear *human signal* mandatory, but meaningless. As noted, it thwarted the scientific method.

The manipulation and politics didn't stop there. The Technical Reports of the three IPCC Working Groups are set aside while another group prepares the Summary for Policymakers (SPM). A few of the scientists prepare a first draft for review by governments. They produce a second draft, then a final report as a compromise between scientists and the individual government representatives. Some claim the scientists set the final summary content, but governments set the form. They release the SPM at before the Science Report. Most of the scientists involved in the science report see the Summary for the first time when it is released to the public. The time between its release to the public and the release of the Technical Report allows time to make sure it aligns with what the politicians/scientists have concluded.

Here is the instruction in the IPCC procedures:

Changes (other than grammatical or minor editorial changes) made after acceptance by the Working Group or the Panel shall be those necessary to ensure consistency with the Summary for Policymakers (SPM) or the Overview Chapter.[68]

Yes, you read that correctly. It is like an Executive writing a summary and then having employees write a report that agrees with the summary.

When you accept the hypothesis before it is proven, you step on the treadmill of maintaining the hypothesis. This leads to selective and biased research and publications. As evidence appeared to show problems with the hypothesis, the natural tendency was to become more vigorous in defending the increasingly indefensible. John Maynard Keynes' sardonic question underlines the human tendency:

If the facts change, I'll change my opinion. What do you do, Sir?

The IPCC and those who participate produce results predetermined to a conclusion by the rules, regulations and procedures carefully crafted by Maurice Strong. These predetermined the outcome—a situation in complete contradiction to the objectives and methods of science.

As evidence grew that the hypothesis was scientifically unsupportable, adherents began to defend rather than accept and adjust. All efforts were in a constant search for an identifiable human signal. Next we will examine how the political system Strong and the UN he organized allowed perpetuation of incorrect science and falsely identified smoking guns.

---

[68] http://www.ipcc.ch/pdf/ipcc-principles/ipcc-principles-appendix-a.pdf

BEN SANTER

CHAPTER SIX

Nature Fails to Cooperate with the IPCC Deception

The Paradox

PEOPLE THINK ALL NATIONAL policies and public understanding of global warming and climate are based on all three IPCC Reports. In reality they use only *The Summary for Policymakers* (SPM).[69] If they used the Science Report titled *The Physical Science Basis* of Working Group I, understanding and policies would be entirely different. Indeed, global warming and climate change would not be an issue. Actually, remarkably few read the Science Report. It presents an altogether different picture.

The Science Report is full of cautions, warnings and clear statements of the limitations; the message is implicit (explicit) that IPCC science is not an adequate basis for policy. The SPM is the

---

69

http://www.ipcc.ch/publications_and_data/ar4/wg1/en/contents.html

antithesis, replete with unjustified certainties and urgent expressions of the need for action.

None of this is coincidental. It was the objective of the IPCC to create scientific proof for a political objective. It was frighteningly successful, but constantly faced with the inadequacies of the science and the failure of nature to cooperate. By relating the problems they could later say they knew the limits. It will be the final deception as the mainstream media and the public start to catch on. They will be able to say the media and public failed; there was no attempt to deceive.

A Structure to Achieve the Predetermined Goal

Structure and procedures of the IPCC were designed to prove that human production of CO_2 was causing global warming/climate change; what became known as the Anthropogenic Global Warming (AGW) theory. Science works by creating theories based on assumptions, then other scientists—performing their sceptical role—test them. The structure and mandate of the IPCC was in direct contradiction to this scientific method. They set out to prove the theory rather than disprove it. They did not do what Karl Popper said was necessary, namely falsify the hypothesis.[70] They deliberately thwarted the scientific method.

Roy Spencer explained the objective from an inside perspective when he wrote:

> *Politicians formed the IPCC over 20 years ago with an endgame in mind: to regulate CO_2 emissions. I know, because I witnessed some of the behind-the-scenes planning. It is not a scientific organization. It was organized to use the government-funded scientific*

---

[70]http://www.stephenjaygould.org/ctrl/popper_falsification.html

research establishment to achieve a policy goal.[71]

As we have seen, chief architect Maurice Strong made sure the focus was on a human cause of change with CO_2 as the primary culprit. He held that the industrialized nations were destroying the planet through runaway global warming. The first major bias was a very narrow definition of climate change from article 1 of the UNFCCC. It made the human impact the primary purpose of the research. Properly, a scientific definition would put natural climate variability first, but the UNFCCC definition predetermined how the research and results would be narrow and political. It made discovering a clear *human signal* mandatory, but that is impossible unless the amount and cause of natural climate change is known.

Other parts of their mandate illustrate the political nature of the entire exercise. Its own principles require that the IPCC:

> *...shall concentrate its activities on the tasks allotted to it by the relevant WMO Executive Council and UNEP Governing Council resolutions and decisions as well as on actions in support of the UN Framework Convention on Climate Change.*[72]

The role is also to:

> *...assess on a comprehensive, objective, open and transparent basis the scientific, technical and socio-economic information relevant to understanding the*

---

[71] http://www.drroyspencer.com/2011/02/on-the-house-vote-to-defund-the-ipcc/

[72] From: Principles Governing IPCC work, approved at the 14th Session, Vienna 1-3 October 1998 and amended at the 21st Session, Vienna 6-7 November, 2003.

> *scientific basis of risk of human-induced climate change,*
> *its potential impacts and options for adaptation and*
> *mitigation. IPCC reports should be neutral with respect to*
> *policy...*[73]

The cynicism of this last sentence is that they create the Summary for Policymakers (SPM), the most important part of IPCC reports, and these were far from neutral.

Reach a Conclusion Then Make the Research Fit

They guaranteed pre-eminence of the political message over the science by releasing the Summary for Policymakers before the Science Report. They also required the Science Report agree with the Summary when it should be the other way round. This is why there is a delay between the SPM release and the Science Report release.

Warnings of the dangers in this process appeared early, but were easily suppressed. Basing Agenda 21 on climate and the environment gave them the moral high ground, which they used to control and centralize power. It gave them a missionary zeal that colored their actions and activities—achieving the goal at almost any cost. Vaclav Klaus identified this in his book *Blue Planet in Green Shackles* when he wrote:

> *Today's debate about global warming is essentially a*
> *debate about freedom. The environmentalists would like*
> *to mastermind each and every possible (and impossible)*
> *aspect of our lives.*

It is likely that Agenda 21 is "the cause" discussed in the leaked CRU emails.

---

[73] Ibid.

I can't overstate the HUGE amount of political interest in the project as a message that the Government can give on climate change to help them tell their story. They want the story to be a very strong one and don't want to be made to look foolish.

Peter Thorne sensed what was happening and issued a warning:

I also think the science is being manipulated to put a political spin on it which for all our sakes might not be too clever in the long run.

Overall attitude given by the comments and actions suggest the end justifies the means. They were likely emboldened by the demonstrated ability to protect scientists who acted rashly for the cause, such as the now infamous "Chapter 8" fiasco.

The first action that publicly exposed the modus operandi occurred with the 1995 second IPCC Report. It involved Benjamin Santer, a CRU graduate, where he completed a PhD under the supervision of Director Tom Wigley. The thesis, titled *Regional Validation of General Circulation Models,* used three top computer models to recreate North Atlantic conditions. The area chosen underscores the most serious limitation in climate modeling—the lack of data where data was sorely needed. The models created massive pressure systems that don't exist in reality—so Santer knew the limitations of the models.

Appointed lead-author for Chapter 8 of the 1995 IPCC Report titled *Detection of Climate Change and Attribution of Causes,* Santer apparently determined to prove humans were a factor by altering the meaning of what was agreed by the others at the draft meeting in Madrid. It is notable how many of the people appointed to key positions on the IPCC were new young graduates who were guided by academic mentors or were saying what was required to perpetuate the false science.

Chapter Six

To the consternation of many, Dr Frederick Seitz and Dr. Fred Singer disclosed how Santer made changes ostensibly to accommodate the SPM rule cited earlier.

However, as Seitz wrote in reference to the 1995 report:

> *I have never before witnessed a more disturbing corruption of the peer-review process than the events that led to this IPCC report.*

Here are the comments agreed to by all authors of the Chapter prior to their inclusion in the Summary for Policymakers:

> *None of the studies cited above has shown clear evidence that we can attribute the observed [climate] changes to the specific cause of increases in greenhouse gases.*
>
> *While some of the pattern-base discussed here have claimed detection of a significant climate change, no study to date has positively attributed all or part of climate change observed to man-made causes.*
>
> *Any claims of positive detection and attribution of significant climate change are likely to remain controversial until uncertainties in the total natural variability of the climate system are reduced.*
>
> *While none of these studies has specifically considered the attribution issue, they often draw some attribution conclusions, for which there is little justification.*

Here are Santer's replacements:

> *1. There is evidence of an emerging pattern of climate response to forcing by greenhouse gases and sulfate aerosols...from the geographical, seasonal and vertical patterns of temperature change...These results point*

toward a human influence on global climate.

2. The body of statistical evidence in chapter 8, when examined in the context of our physical understanding of the climate system, now points to a discernible human influence on the global climate...

Notice this is "statistical evidence", not actual evidence, but is part of the desire to blame humans. Compare it with the comment in the 1990 IPCC report:

...it is not possible at this time to attribute all, or even a large part, of the observed global-mean warming to [an] enhanced greenhouse effect on the basis of the observational data currently available.

The issue didn't change in the five years between IPCC reports, but that wasn't what was needed.

As Avery and Singer noted in 2006:

Santer single-handedly reversed the 'climate science' of the whole IPCC report and with it the global warming political process! The 'discernible human influence' supposedly revealed by the IPCC has been cited thousands of times since in media around the world, and has been the 'stopper' in millions of debates among nonscientists.[74]

A quick cover-up was necessary so the procedure, that was to appear often in the future, was applied.

---

[74] *Unstoppable Global Warming: Every 1,500 Years, Updated and Expanded Edition*, 2007, S. Fred Singer and Dennis T. Avery, Rowman & Littlefield, p. 64.

On July 4th, 1996, the compliant and often complicit journal *Nature* conveniently published *A Search for Human Influences on the Thermal Structure of the Atmosphere* with a familiar list of authors: Santer, Wigley, Jones, Mitchell, Oort and Stouffer. It provided observational evidence that proved the models were accurate. A graph is worth a thousand words and so it was with Santer's "discernible human influence." John Daly recreated Santer et al's graph[75] (see Figure 2) of the upward temperature trend in the Upper Atmosphere.

Errors were spotted quickly, one identified the cherry picking, the other a natural explanation for the pattern, but *Nature* didn't publish the rebuttals until five months later on Dec 12, 1996. By that time the PR cover-up was under way. On July 25th, 1996, the American Meteorological Society (AMS) sent a letter of defense to Santer. The letter appears to be evidence of CRU influence and a PR masterpiece. It said there were two questions, the science, and what society must do about scientific findings and the debate they engendered. Science should only be debated in *peer-reviewed scientific publications—not the media*. This was part of the strategy of controlling peer review and dismissing accurate criticisms from bloggers confirmed in a leaked email from Michael Mann:

> *This was the danger of always criticizing [sic] the skeptics for not publishing in the "peer-reviewed literature". Obviously, they found a solution to that— take over a journal!*[76]

The problem of peer review was also misrepresented. As Singer and Avery observe in their book *Unstoppable Global Warming*:

---

[75] Figure 1 at http://www.john-daly.com/sonde.htm
[76] http://www.ecowho.com/foia.php?file=1047388489.txt&search=Hans+von+Storch

The IPCC's defenders claim that the crucial chapter 8 of the panel's Climate Change 1995 was based on 130 peer-reviewed science studies. Actually, the chapter was based mainly on two research papers by its lead author, Ben Santer, of the U.S. government's Lawrence Livermore National Laboratory. Neither of the Santer papers had been published at the time the chapter was under review and they had not been subject to peer review. Scientific reviewers subsequently learned that both the Santer papers shared the same defect as the IPCC's chapter 8: Their "linear upward trend" occurs only from 1943 to 1970. [See Figure 2]...

Then AMS wrote:

What is important scientific information and how it is interpreted in the policy debates is an important part of our jobs. That is, after all, the very reasons for the mix of science and policy in the IPCC.

Daly correctly called this *Scientism*. Santer reportedly later admitted:

...he deleted sections of the IPCC chapter which stated that humans were not responsible for climate change.[77]

He did not admit the changes at the time and achieved the objective of getting the "discernible human influence" message on the world stage. He was protected by the group that demonstrated its control over peer review, journals, professional societies, and the media, that was disclosed in the emails leaked

---

[77] http://www.prisonplanet.com/exclusive-lead-author-admits-deleting-inconvenient-opinions-from-ipcc-report.html

from the CRU in November 2009 and reinforced in 2011.

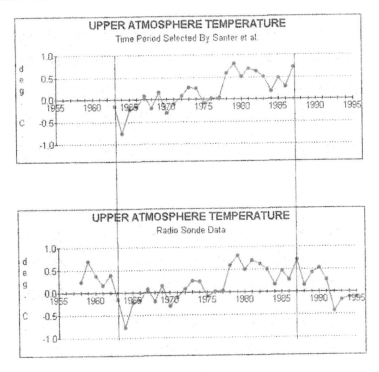

Figure 2

Daly's Identification of Santer's Selected Portion of the Record[78]

The issue that humans were causing global warming was key to the entire objective of the founders and orchestrators of the IPCC.

Seitz's comments couldn't go unchallenged—he was almost immediately attacked as the Marshall Institute described:

> *The campaign of personal destruction propagated by environmental advocacy groups hit a new low with the*

---

[78] http://www.john-daly.com/sonde.htm

release of the May issue of Vanity Fair and the
subsequent press conference today by the National
Environmental Trust. Their accusations about Dr.
Frederick Seitz are unfounded and unbelievable. It
reflects a campaign of character assassination and the
lack of any standards of decency.

The manipulation of the reports was to continue and get worse, as were the attacks on those who questioned the established narrative.

No single event symbolizes the political focus and manipulations to achieve the goal of identifying humans as the cause of climate change, but the Chapter 8 fiasco was the first to reach the world stage. Reaction to its exposure brought the now familiar pattern of responses. These include strident denials of any agenda by suggesting those who raised the questions had the agenda. Outrage at the suggestion a poor hard working scientist would have a political agenda. Personal attacks on the individuals who asked questions including innuendoes and direct charges of their associations and challenges of their general credentials especially climate qualifications, and most important of all, funding.

The climate debate was now a purely political battle. Science was increasingly and rudely pushed aside. The misdirection and machinations within the IPCC were to get worse.

The powers in charge, including Sir John Houghton, co-chair of the IPCC at the time, were able to overcome the scandal mostly because they could claim it was a minor isolated error. In retrospect, it was crucial as it appears Santer carried out his intended role clumsily as a new over-eager participant in the IPCC process. It was the first of many deceptive false claims, and manipulation necessary to prove a human cause that continued despite the Chapter 8 warnings. The most egregious was to come later in the 2001 IPCC Report with the now infamous "hockey stick". What did take hold was the phrase, "a discernible human influence on the global climate," even though it was incorrect.

The extent of IPCC participant's control of the process and apparent feelings of invulnerability pervade the emails leaked from the Climatic Research Unit (CRU) in November of 2009. Many people don't know that control of the IPCC from critical chapters, like data, computer models and paleoclimate, to the SPM were always in the hands of members of the CRU and their affiliates.

By the time of the 2001 IPCC Third Assessment Report (TAR), the politics and hysteria about climate change had risen to a level that demanded clear evidence of a human signal. An entire industry had developed round massive funding from government. A large number of academic, political and bureaucratic careers had evolved and depended on expansion of the evidence.

Environmentalists were increasing pressure on the public and thereby politicians. In addition, they lifted the bar of proof by claiming the 20th century and especially the last decade had nine of the ten warmest years in history, warming beyond anything previous and therefore unnatural. These claims were to become their downfall because, as some climate experts knew, there were much warmer periods in the historic record.

Most troubling for the IPCC was a temperature curve produced by Professor Lamb included in the first IPCC Report (Figure 3). It showed the Medieval Warm Period (MWP) followed by a cool period called the Little Ice Age (LIA).

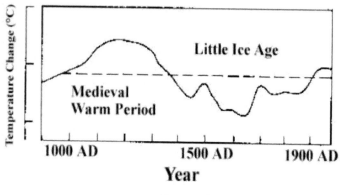

Figure 3

Schematic Diagram of Global Temperature Variations since the Pleistocene—the Last 1,000 Years[79]

There were hundreds of research papers from a wide variety of sources confirming the existence of a period warmer than today just a thousand years ago known as the MWP. The work of Soon and Baliunas documented its existence.[80]

This period was warmer than modern temperatures and warmer than some computer model predictions for the future. Its existence was a serious problem because it negated the claims that the 20th century temperatures were unprecedented. Phil Jones flagged the problems created by Lamb's graph produced as Figure 7c in early IPCC reports. He advised them not to discuss it openly at RealClimate:

> *Keep the attached to yourself. I wrote this yesterday, but still need to do a lot more. I added in a section about*

---

[79]http://www.ipcc.ch/ipccreports/far/wg_I/ipcc_far_wg_I_full_rep ort.pdf, p. 202

[80] *Proxy climatic and environmental changes of the past 1,000 years*, Soon, W., and S. Baliunas, 2003 Climate Research, 23, 89–110.

post-Lamb work in CRU, but need to check out the references I've added and look at the extra one from 1981 that you've sent. This may take me a little time as I'm away Weds/Thurs this week. I see my name on an abstract, by the way, that I have no recollection of! I presume this has something in about instrumental global temps. This abstract isn't in my CV!!!!!

So your point (3) needs to document that we knew the diagram wasn't any good, as well as how far back it goes. Knowing Hubert on some of his other 'breakthroughs!' it is clearly possible it goes back to Brooks!

On the post-Lamb work in CRU, I recall talking to Graham (maybe mid-1980s) when he was comparing recent CRU work with Lamb—correlations etc. Did that ever see the light of day in these pubs or elsewhere? I will look. It isn't in the chapter Astrid and he wrote in the CRU book from 1997. I recall some very low correlations—for periods from 1100 to 1500.

This is all getting quite complex. It clearly isn't something that should be discussed online on RC—at least till we know all the detail and have got the history right as best we can. A lot of this history is likely best left buried, but I hope to summarize enough to avoid all the skeptics wanting copies of these non-mainstream papers. Finding them in CRU may be difficult!

As for who put the curve in—I think I know who did it. Chris may be ignorant of the subject, but I think all he did was use the DoE curve. This is likely bad enough. I don't think it is going to help getting the real culprit to admit putting it together, so I reckon Chris is going to get the blame.

I have a long email from him—just arrived. Just read that and he seems to changing his story from last

December, but I still think he just used the diagram.
Something else happened on Friday—that I think put me
onto a different track. This is all like a mystery
whodunit.[81]

The First Attack

There were two attacks. One still unknown to many and unreported by the mainstream media involved denigration and vicious attacks on Soon and Baliunas. Maybe it went unreported because it involved, in a truculent nasty manner, a member of the Obama White House.

There is a multitude of small but disturbing stories in the extensive leaked email files. For example, I've known solar physicists Sallie Baliunas and Willie Soon for a long time. I published an article with Willie and enjoyed extensive communication. I was on advisory committees with Sallie when she suddenly and politely withdrew from the fray. I don't know if the following events were contributing factors, but it seems likely.

Baliunas and Soon excellent work confirmed the existence of the Medieval Warm Period (MWP) from a multitude of sources. It challenged attempts to get rid of the MWP because it contradicted the claim by the proponents of anthropogenic global warming (AGW). Several scientists challenged the claim that the latter part of the 20th century was the warmest ever. They knew the claim was false because many warmer periods occurred in the past. Michael Mann *got rid* of the MWP with his production of the hockey stick, but Soon and Baliunas were problematic. What better than to have a powerful academic destroy their credibility for you? Sadly, there are always people who will do the dirty work.

---

[81] http://www.telusplanet.net/dgarneau/climate-e-mails.htm

Chapter Six

A perfect person and opportunity appeared. On October 16th, 2003 Michael Mann sent an email to people involved in the CRU scandal:

Dear All,
Thought you would be interested in this exchange, which
John Holdren of Harvard has been kind enough to pass
along...[82]

At the time, Holdren was Teresa and John Heinz Professor of Environmental Policy & Director, Program in Science, Technology, & Public Policy, Belfer Center for Science and International Affairs, John F. Kennedy School of Government. Later he became Director of the White House Office of Science and Technology Policy, Assistant to the President for Science and Technology, and Co-Chair of the President's Council of Advisors on Science and Technology—informally known as the United States Science Czar.

In an email on October 16th, 2003 from John Holdren to Michael Mann and Tom Wigley we're told:

I'm forwarding for your entertainment an exchange that
followed from my being quoted in the Harvard Crimson
to the effect that you and your colleagues are right and
my "Harvard" colleagues Soon and Baliunas are wrong
about what the evidence shows concerning surface
temperatures over the past millennium. The cover note to
faculty and postdocs in a regular Wednesday breakfast
discussion group on environmental science and public
policy in Harvard's Department of Earth and Planetary
Sciences is more or less self-explanatory.

---

[82] Climategate email from Michael Mann, October 16th, 2003

73

This is what Holdren sent to the Wednesday Breakfast group:

> *I append here an e-mail correspondence I have engaged in over the past few days trying to educate a Soon/Baliunas supporter who originally wrote to me asking how I could think that Soon and Baliunas are wrong and Mann et al. are right (a view attributed to me, correctly, in the Harvard Crimson). This individual apparently runs a web site on which he had been touting the Soon/Baliunas position.*

The exchange Holdren refers to is a challenge by Nick Schulz editor of Tech Central Station (TCS). On August 9th, 2003 Schulz wrote:

> *In a recent Crimson story on the work of Soon and Baliunas, who have written for my website[83], you are quoted as saying: My impression is that the critics are right. It is unfortunate that so much attention is paid to a flawed analysis, but that's what happens when something happens to support the political climate in Washington. Do you feel the same way about the work of Mann et. al.? If not why not?*

Holdren provides lengthy responses on October 13th, 14th, and 16th, but his comments fail to answer Schulz's questions. After the first response Schulz replies:

> *I guess my problem concerns what lawyers call the burden of proof. The burden weighs heavily much more heavily, given the claims on Mann et.al. than it does on Soon/Baliunas. Would you agree?*

---

[83] www.techcentralstation.com

Of course, Holdren doesn't agree. He replies:

> *But, in practice, burden of proof is an evolving thing—it evolves as the amount of evidence relevant to a particular proposition grows.*

No, it doesn't evolve; it is either on one side or the other. This argument is in line with what has happened with AGW. He then demonstrates his lack of understanding of science and climate science by opting for Mann and his hockey stick over Soon and Baliunas. His entire defense and position devolves to a political position. His attempt to belittle Soon and Baliunas in front of colleagues is a sad measure of the man's character.

Schulz provides a solid summary when he writes:

> *I'll close by saying I'm willing to admit that, as someone lacking a PhD, I could be punching above my weight. But I will ask you a different but related question. How much hope is there for reaching reasonable public policy decisions that affect the lives of millions if the science upon which those decisions must be made is said to be by definition beyond the reach of those people?*

We now know they deliberately placed it beyond the reach of the people and restricted it to the group that he used to ridicule Soon and Baliunas. It appears he was blinded by his political views, which are central to the Club of Rome theme of overpopulation, and as his record shows, are frightening. One web site synthesizes his position as follows:

> *Forced abortions. Mass sterilization. A "Planetary Regime" with the power of life and death over American*

citizens.[84]

The Second Attack

The second attack was on Lamb's graph. People find it hard to believe that scientists involved with the IPCC and mostly affiliated with the Climatic Research Unit (CRU) set out to rewrite history. But that is exactly what they did. Professor Deming revealed the evidence in the following letter to *Science*:

> *With the publication of the article in Science [in 1995], I gained significant credibility in the community of scientists working on climate change. They thought I was one of them, someone who would pervert science in the service of social and political causes. So one of them let his guard down. A major person working in the area of climate change and global warming sent me an astonishing email that said, **"We have to get rid of the Medieval Warm Period."** [Emphasis added]*

They did this with what became known as the *hockey stick*. The name came from the shape of a graph, which showed no temperature increase for 1000 years (the handle) with a sudden rise in the 20th century (the blade) (Figure 4).

It was an ideal solution, because it was by design. It eliminated the MWP and the LIA and introduced the sudden rise in the 20th century as apparently unnatural. It had to be due to human activity; it reinforced Santer's claim of a "clearly discernible human signal."

---

[84] http://zombietime.com/john_holdren/

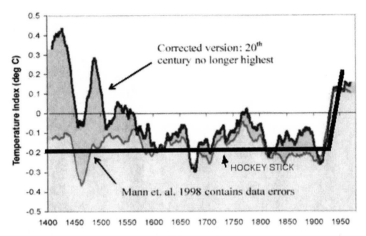

Figure 4

After McIntyre and McKitrick with Hockey Stick Superimposed by Author

Research that produced the hockey stick came primarily from dendroclimatology: the reconstruction of past climates from tree ring data. Their problem was the tree ring data showed temperature declining in the 20th century. The solution was to drop the latter part of the tree ring record and tack on modern temperature data to create the blade. This is the now infamous "Mike's Nature trick" designed to "hide the decline".

They incorrectly assumed tree rings are only a function of temperature and cherry-picked those trees that gave the desired result. When challenged on this one dendrochronologist justified this practice by telling a U.S. Congressional committee:

> *You have to pick cherries if you are going to make cherry pie.*[85]

---

[85] Rosanne D'Arrigo, Tree-Ring Laboratory, Lamont-Doherty Earth Observatory of Columbia University addressing a National Academy of

Another wrote:

> However as we mentioned earlier on the subject of biological growth populations, this does not mean that one could not improve a chronology by reducing the number of series used if the purpose of removing samples is to enhance a desired signal. The ability to pick and choose which samples to use is an advantage unique to dendroclimatology.[86]

These are truly disturbing comments in any area of research.

A dendroclimatic study published in 1998 by Mann, Bradley and Hughes, (known as MBH98) and introduced in Chapter 2 of the Technical Report (produced by Working Group I) became the centre of controversy. Conflict screamed, because Mann was a lead author of the Chapter while Bradley and Hughes were contributing authors, but it was ignored. It screamed louder when the hockey stick appeared as a major part of the Summary for Policymakers for which Mann was a contributing author. It followed an opening statement that said:

> New analyses of proxy data for the Northern Hemisphere indicate that the increase in temperature in the 20th century is likely to have been the largest of any century during the past 1,000 years. It is also likely that, in the Northern Hemisphere, the 1990s was the warmest decade and 1998 the warmest year.[87]

Sciences Panel Hearing in 2006 as referenced by Steve McIntyre, http://climateaudit.org/2006/03/07/darrigo-making-cherry-pie/
[86] http://climateaudit.org/2006/03/09/esper-on-in-site-cherry-picking/
[87] http://www.ipcc.ch/ipccreports/tar/wg1/005.htm

The graph appeared on the second page of the Summary so it quickly, visually and scientifically underscored the argument. It also—as intended—stole the media limelight. Versions quickly appeared in everything from *National Geographic*, to Al Gore's movie,[88] to government web sites. Now they could bully people who questioned the science and introduce draconian legislation to get rid of the evil CO_2 as was the intention all along. Now the useless Kyoto Protocol apparently had its justification.

The hockey stick fiasco failed a basic scientific test, known as reproducible results. Other scientists use the same data and procedures to try and reproduce the original findings. Steve McIntyre and Ross McKitrick (MM) attempted, but failed to reproduce the MBH98 findings. A debate ensued with claims MM were wrong or not qualified climate experts. They replied that Mann had refused to disclose all the codes he used to achieve the results, but even without them the main problem was a misuse of data and statistical techniques. In effect, the hockey stick was meaningless.

The U.S. National Academy of Sciences (NAS) appointed a committee chaired by Professor Wegman to investigate and arbitrate. His committee report found in favor of MM as follows:

> *It is not clear that Mann and associates realized the error in their methodology at the time of publication. Because of the lack of full documentation of their data and computer code, we have not been able to reproduce their research. We did, however, successfully recapture similar results to those of MM. This recreation supports the critique of the MBH98 methods, as the offset of the mean value creates an artificially large deviation from the desired mean value of zero.*[89]

---

[88] http://www.imdb.com/title/tt0497116/

[89] http://www.uoguelph.ca/~rmckitri/research/WegmanReport.pdf

Mann continues to withhold critical data that precludes completion of the basic scientific test of reproducible results. He and his acolytes still fight a rearguard action claiming their work was valid.

Several people raised serious concerns about the objectivity of an IPCC Report and Summary with substantial input from scientists citing their own research. Unfortunately, this is typical of the incestuous political nature of the entire IPCC process. In his report Professor Wegman's first recommendation says:

> Especially when massive amounts of public monies and human lives are at stake, academic work should have a more intense level of scrutiny and review. It is especially the case that authors of policy-related documents like the IPCC report, Climate Change 2001: The Scientific Basis, should not be the same people as those that constructed the academic papers.[90]

Most people, especially the media, missed the equally startling and disturbing conclusion by Wegman:

> In our further exploration of the social network of authorships in temperature reconstruction, we found that at least 43 authors have direct ties to Dr. Mann by virtue of coauthored papers with him. Our findings from this analysis suggest that authors in the area of paleoclimate studies are closely connected and thus 'independent studies' may not be as independent as they might appear on the surface.[91]

The incestuous potentials of such a small close-knit group are

---

[90] Ibid.
[91] Ibid.

disturbing beyond co-authorship. Proponents of the anthropogenic global warming theory made much of the fact that critics have few or no 'peer reviewed' papers.

Why?

It appears members of the group of 43 were also peer reviewing each other's papers. It is one possible explanation why Mann's paper sailed through peer review. Journal editors are not required to disclose the names of reviewers, so we can't know. It probably also explains why so much is made of peer review by members and defenders of the IPCC. When there is a small group in a specialized research area, it is too easy to control what gets published. It is what I call peer review censorship.

CRU emails reveal the extent of attempts to block publication of McIntyre and McKitrick's analysis of the hockey stick graph. Using RealClimate[92], Mann and Schmidt launched a PR attack on the peer review process that approved publication of MM's first article. They referenced an article 'in press' by Rutherford et al with the *Journal of Climate*, but Mann was one of the authors. Equally troubling is the editorship of that Journal. Donna Laframboise writes:

> *We're supposed to trust the conclusions of the Intergovernmental Panel on Climate Change (IPCC) because much of the research on which it relies was published in peer-reviewed scientific journals. But what happens when the people who are in charge of these journals are the same ones who write IPCC reports?[93]*

She identifies 14 people involved in the Journal and the IPCC,

---

[92]http://www.realclimate.org/index.php/archives/2005/01/peer-review-ii/

[93] http://nofrakkingconsensus.com/2011/08/23/the-journal-of-climate-the-ipcc/

including Michael Mann and Gavin Schmidt. Laframboise asks this question: *Is this not too incestuous for words?*

Does this accusation apply to the fact that CRU/IPCC luminary, including Mann, dominate the list of people involved in the *Journal of Climate*?[94] It is just one part of how a small group could control a scientific issue for a political agenda.

The Deception: Make an Unproven Theory Become Fact

Starting in 1990, the IPCC produced Summaries with each increasing the probability of their claim of a clearly discernible human signal.

Despite evidence discrediting claim after claim, the process did not stop. It continued with the 2013 Assessment Report 5 (AR5). In the face of failed predictions (projections) they increased their claim of certainty in their results from >90% to 95%. It continues the pattern of ignoring the obvious—there are none so blind as those who will not see.

Throughout IPCC history a combination of events drove the misconceptions forward—overriding any attempts to point out errors, omissions and deliberate deceptions. Mainstream media bought into and promoted the unproven theory. Massive amounts of government funding went through the WMO national agencies to support research that proved the theory. Scientists who challenged were denied funding and marginalized. National environmental policies were introduced based on the misleading information of the Summaries.

Very few people read IPCC Science Reports, and most wouldn't understand them if they did. This is especially true of the chapter on climate models. However, it is the most critical

---

[94] http://nofrakkingconsensus.com/2011/08/23/the-journal-of-climate-the-ipcc/

area to IPCC claims that human CO_2 is causing warming. They produce the predictions used to threaten impending doom. Yet, despite their failures, they retain not so much credibility as awe.

Limitations of the models were manifest in the switch from making forecasts to producing scenarios; a move forced by the consistent failure of the forecasts as early as the 1995 Report.

Despite the switch, the public continues to believe they were making forecasts and IPCC scientists did nothing to disabuse them. This is a vital part of the deception. In the Science Report produced by Group I they provide a range of temperature scenarios. Of course, the sensationalist media always focus on the highest numbers, implying they are predictions. The deception is part of the entire pattern of IPCC behaviour and once again occurs in the gap between the Science report and the Summary for Policymakers. The latter specifically says they are predictions in several places:

> Based on current models we predict:[95]

and

> Confidence in predictions from climate models[96]

Nowhere is the difference more emphasized than in this comment from the Science Report in TAR:

> In climate research and modelling, we should recognize that we are dealing with a coupled non-linear chaotic system, and therefore that the long-term prediction of future climate states is not possible.

---

[95] IPCC, FAR, SPM, p. xi
[96] IPCC, FAR, SPM, p. xxviii

Now you know why the IPCC made sure they released the Summary first.

World policy based on the output of the IPCC models is already bringing disasters. We are told model reliability was poor, but has improved. They have not improved in 18 years. Harrabin's comment about the Reading Conference says:

> So far modellers have failed to narrow the total bands of uncertainties since the first report of the Intergovernmental Panel on Climate Change (IPCC) in 1990.[97]

Koutsoyiannis et al confirmed this in April 2008 where in an article they found:

> The GCM (General Circulation Models) outputs of AR4 (FAR) as compared to those of TAR, are a regression in terms of the elements of falsifiability they provide...[98]

Most devastating is their finding: *This makes the future climate projections not credible*, but this is not surprising given even a very brief examination of the models. Roger Harrabin, a BBC reporter, quotes Professor Roy Spencer saying:

> He thinks clouds are impossible to model at present.

97

http://www.theregister.co.uk/Print/2009/06/24/met_office_ukcp/
[98] Koutsoyiannis, D., N. Mamassis, A. Christofides, A. Efstratiadis and S.M. Papalexiou, *Assessment of the reliability of climate predictions based on comparisons with historical time series,* European Geosciences Union General Assembly 2008, Geophysical Research Abstracts, Vol. 10, Vienna, 09074, European Geosciences Union, 2008.

Chapter Six

Imagine using a global climate model that cannot handle clouds as the basis for policy to devastate world economies.

Meanwhile, the IPCC violated many scientific principles. As Green and Armstrong note:

> We audited the forecasting processes described in Chapter 8 of the IPCC's WG1 Report to assess the extent to which they complied with forecasting principles. We found enough information to make judgments on 89 out of a total of 140 forecasting principles. The forecasting procedures that were described violated 72 principles. Many of the violations were, by themselves, critical...The forecasts in the Report were not the outcome of scientific procedures. In effect, they were the opinions of scientists transformed by mathematics and obscured by complex writing. Research on forecasting has shown that experts' predictions are not useful. We have been unable to identify any scientific forecasts of global warming. Claims that the Earth will get warmer have no more credence than saying that it will get colder.

Ironically, the IPCC model forecasts are also their weakness because the public are very aware of weather forecast failures. AGW proponents tried to deflect the cynicism by saying there was a difference between weather and climate forecasts. The problem is climate is the average weather, so the mechanisms are the same. For example, Dr Donald Dubois noted:

> If the major climate models that are having a major impact on public policy were documented and put in the public domain, other qualified professionals around the world would be interested in looking into the validity of these models. Right now, climate science is a black box that is highly questionable with unstated assumptions

85

and model inputs.[99]

Or, as Dr. David Wojick explained:

> *The public is not well served by this constant drumbeat of false alarms fed by computer models manipulated by advocates.*[100]

New Scientist reported that Tim Palmer, a leading climate modeller at the European Centre for Medium-Range Weather Forecasts in Reading England said:

> *I don't want to undermine the IPCC, but the forecasts, especially for regional climate change, are immensely uncertain.*[101]

Additionally, they report that he speaks to the discrepancy between faith in IPCC Reports and reality:

> *...he does not doubt that the Intergovernmental Panel on Climate Change (IPCC) has done a good job alerting the world to the problem of global climate change. But he and his fellow climate scientists are acutely aware that the IPCC's predictions of how the global change will affect local climates are little more than guesswork. They*

---

[99]

http://www.naturalclimatechange.us/natural%20climate%20change-%20adds%2012-1-10/new%20power%20pts.-%20AGW%20quotes/AGW%20quotes%20on%20IPCC%20fraud.ppt x

[100] http://forums.poz.com/index.php?topic=17918.25;wap2

[101] http://www.newscientist.com/article/mg19826543.700-poor-forecasting-undermines-climate-debate.html

> *fear that if the IPCC's predictions turn out to be wrong,*
> *it will provoke a crisis in confidence that undermines the*
> *whole climate change debate.*[102]

The crisis in confidence about the climate debate is good news. However, a growing problem is the loss of credibility for science on all issues, especially the environment.

Emeritus Professor Garth Paltridge explains what is happening:

> *Basically, the problem is that the research community has gone so far along the path of frightening the life out of the man in the street that to recant publicly even part of the story would massively damage the reputation and political clout of science in general. And so, like corpuscles in the blood, researchers all over the world now rush in overwhelming numbers to repel infection by any idea that threatens the carefully cultivated belief in climatic disaster.*[103]

We now know, through leaked emails from the Climatic Research Unit (CRU) at the University of East Anglia (UEA), how a small group who were also members of the IPCC, created a totally false picture supposedly based on science. Some have described what the IPCC achieved as similar to Lysenkoism. *The Skeptics Dictionary* explains:

> *Under Lysenko's guidance, science was guided not by the most likely theories, backed by appropriately controlled experiments, but by the desired ideology. Science was practiced in the service of the State, or more precisely, in*

---

[102] Ibid.

[103] http://archive-jer.blogspot.com/2009/03/to-tell-truth_14.html

the service of ideology.[104]

Lysenko's version of genetics dominated and seriously diverted Soviet science from 1948 to 1965 until finally rejected. Certainly the concept that human CO_2 causes warming and climate change was based on unproven theory used by people with an ideology. They used instruments of state to dominate the science. They also attacked and abused anyone who dared to pursue proper science. The small group who controlled the IPCC were unlikely to change their tune. A pattern that was borne out by the release of IPCC Report AR5 in September 2013, which denied the fact that for 17 years global temperature declined slightly while CO_2 levels continued to increase.

The group, led by Canadian Donna Laframboise, was responding to IPPC Chairman Rajendra Pachauri's comment in the *Times of India*:

> *IPCC studies only peer-review science. Let someone publish the data in a decent credible publication. I am sure IPCC would then accept it, otherwise we can just throw it into the dustbin.*[105]

In addition, they found that:

> *21 of 44 chapters in the United Nations' Nobel-winning climate bible earned an F on a report card released today. Forty citizen auditors from 12 countries examined 18,500 sources cited in the report—finding 5,600 to be not peer-reviewed.*

---

[104] http://skepdic.com/lysenko.html
[105] http://nofrakkingconsensus.blogspot.com/2010/04/climate-bible-gets-21-fs-on-report-card.html

KEVIN TRENBERTH

CHAPTER SEVEN

Part One

Scientists Knew from the Start

NORMALLY CREATIVE WRITING instructors advise against using too many quotes. However, when trying to identify that people knew what was going on, it is most effective when it is in their own words. Based on their own comments, most involved with the IPCC knew from the start the limitations of the data and the climate models. Kevin Trenberth, when responding to a report on the inadequacies of the weather data produced by the U.S. National Research Council, said:

> *It's very clear we do not have a climate observing system...this may come as a shock to many people who assume that we do know adequately what's going on with the climate, but we don't.*[106]

---

[106]

http://www.enouranois.gr/english/political/Dialogue_with_a_Climat
e_Change_Contrarian.doc

This was in response to the February 3rd, 1999 report that said:

> *Deficiencies in the accuracy, quality and continuity of the records place serious limitations on the confidence that can be placed in the research results. The problem is we have fewer weather stations and less data now than in 1990.*[107]

A few years ago the Climatic Research Unit (CRU), before it became infamous for leaked emails exposing the manipulation of climate science, issued a statement that said:

> *GCMs are complex, three dimensional computer-based models of the atmospheric circulation. Uncertainties in our understanding of climate processes, the natural variability of the climate, and limitations of the GCMs mean that their results are not definite predictions of climate.*[108]

Phil Jones, Director of the CRU at the time of the leaked emails, and Tom Wigley, a former director of the CRU and player in the IPCC said:

> *Many of the uncertainties surrounding the causes of climate change will never be resolved because the necessary data are lacking.*

Stephen Schneider, a very prominent part of the IPCC from the start, said:

---

[107]

http://news.google.com/newspapers?nid=1915&dat=19990203&id=XgQhAAAAIBAJ&sjid=pXUFAAAAIBAJ&pg=5480,367285

[108] http://www.cru.uea.ac.uk/projects/medalus/gcm.htm

Uncertainty about feedback mechanisms is one reason why the ultimate goal of climate modelling—forecasting reliably the future of key variables such as temperature and rainfall patterns—is not realizable.

Schneider also set the tone when he said in *Discover* magazine:

Scientists need to get some broader based support, to capture the public's imagination...that, of course, entails getting loads of media coverage. So we have to offer up scary scenarios, make simplified dramatic statements, and make little mention of any doubts we may have...each of us has to decide what the right balance is between being effective and being honest.

There is no decision—honesty must **always** trump effectiveness. The IPCC achieved Schneider's objective with dramatic but devastating effect. This is not surprising because the first IPCC Chairman, Sir John Houghton said:

Unless we announce disasters no one will listen.[109]

It appears that the IPCC and its disciples decided that the end justifies the means. Phil Jones, Director of the CRU and major player in the IPCC demonstrated this when on July 8th, 2004 in response to concerns about papers challenging their 'science' he wrote to Michael Mann, with the label "HIGHLY CONFIDENTIAL" that:

I can't see either of these papers being in the next IPCC report. Kevin [Trenberth] and I will keep them out somehow—even if we have to redefine what the peer-

---

[109] http://motls.blogspot.com/2010/02/sir-john-houghton-is-liar.html

reviewed literature is!

They used computer models to produce the results they wanted and bamboozle most people. Even a cursory examination shows they are inadequate, even by the admission of their greatest manipulators, the IPCC.

The Physical Basis of the Models

Figure 5 shows a diagram of how the atmosphere is divided to create climate models.

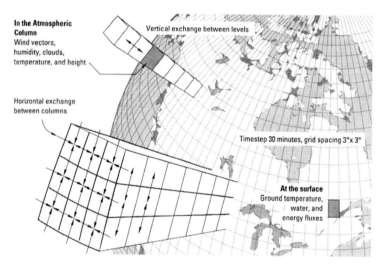

Figure 5

Schematic of General Circulation Model (GCM)

The surface is covered with a grid and the atmosphere is divided into layers. Computer models vary in the size of the grids and the number of layers. The modellers claim a smaller grid provides better results. It doesn't! The model needs data for each cube. However, there are no weather stations for at least 70% of the surface and virtually no data above the surface. There are few records of any length anywhere; the models are built on virtually nothing. In addition, the grid is so large and crude they can't include major weather features like thunderstorms, tornados, or any small storm systems.

They resolve the lack of data using something called parameterization. It is simply very poor estimates produced by computer models. For example, they claim using this method that the weather data from a single station is representative of temperatures for a 1200 km radius circle. Think about that in relation to your community. Also realize this is an example of how they produce data from one model and then use it as real data in another model. I will examine parameterization in more detail later.

O'Keefe and Kueter explain how a model works:

> The climate model is run, using standard numerical modelling techniques, by calculating the changes indicated by the model's equations over a short increment of time—20 minutes in the most advanced GCMs—for one cell, then using the output of that cell as inputs for its neighboring cells. The process is repeated until the change in each cell around the globe has been calculated.[110]

Imagine the number of calculations necessary that even at computer speed of millions of calculations a second takes a long time. The run time is a major limitation. All of this takes huge amounts of computer capacity; running a full-scale GCM for a 100-year projection of future climate requires many months of time on the most advanced supercomputer. As a result, very few full-scale GCM projections are made.

A comment at Steve McIntyre's site, ClimateAudit, illustrates the problem:

> *Caspar Ammann said that GCMs (General Circulation Models) took about 1 day of machine time to cover 25 years. On this basis, it is obviously impossible to model the Pliocene-Pleistocene transition (say the last 2 million years) using a GCM as this would take about 219 years of computer time.*

So you can only run the models if you reduce the number of variables. O'Keefe and Kueter explain:

> *As a result, very few full-scale GCM projections are made. Modellers have developed a variety of short cut techniques to allow them to generate more results. Since the accuracy of full GCM runs is unknown, it is not possible to estimate what impact the use of these short cuts has on the quality of model outputs.*

Omission of variables allows short runs, but allows manipulation and removes the model further from reality. Which variables do you include? For the IPCC only those that create the results they want. Also, every time you run the model it provides a different result because the atmosphere is chaotic. They resolve this by doing several runs and then using an average of the outputs.

By leaving out very important components of the climate system, they increase the likelihood of a human signal being the cause of change. As William Kinninmonth, meteorologist and former head of Australia's National Climate Centre explains:

> *...current climate modeling is essentially to answer one question: how will increased atmospheric concentrations*

of CO_2 (generated from human activity) change earth's temperature and other climatological statistics? Neither cosmology nor vulcanology enter the equations. It should also be noted that observations related to sub-surface ocean circulation (oceanology), the prime source of internal variability, have only recently commenced on a consistent global scale. The bottom line is that IPCC's view of climate has been through a narrow prism. It is heroic to assume that such a view is sufficient basis on which to predict future 'climate'.

The IPCC take the average results of some 22 computers and average them to produce an ensemble result. As Robert Brown, physicist at Duke University notes:

First—and this is a point that is stunningly ignored— there are a lot of different models out there, all supposedly built on top of physics, and yet no two of them give anywhere near the same results!

The title of the article says it all, The "ensemble" of models is completely meaningless, statistically.[111]

Another problem identified with the models was confirmed in a peer review paper published in Monthly Weather Review for July 26th, 2013. It finds that:

...the same global forecast model (one for geopotential height) run on different computer hardware and operating systems produces different results at the output with no other changes.

---

[111] http://wattsupwiththat.com/2013/06/18/the-ensemble-of-models-is-completely-meaningless-statistically/

The problem develops because each model deals with rounding errors differently. These are individually small but because they are numerous the net result is very different outcomes. In the actual studies in the paper they show differences that appear small, until you realize they are for only "10 days worth of modeling." The differences between models are best illustrated graphically as shown in Figure 6 produced by Dr Roy Spencer. [112]

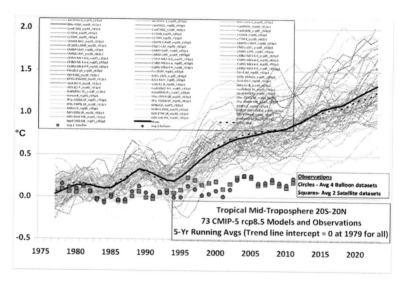

Figure 6

Spencer's Comparison of Computer Model Predictions with Actual Data

---

[112] http://wattsupwiththat.com/2013/07/27/another-uncertainty-for-climate-models-different-results-on-different-computers-using-the-same-code/#more-90513

Chapter Seven

Static Climate Models in a Virtually Unknown Dynamic Atmosphere

Knowledge about the atmosphere and lack of data are serious limitations on understanding climate change and building climate models. The atmosphere is three-dimensional and dynamic, so building a computer model that even approximates reality requires far more data than exists and much greater understanding of an extremely turbulent and complex system.

Sunlight striking the Earth does not impart the same amount of energy everywhere. First, only one half the globe is receiving sunlight at any given time. Second, the amount of energy input is maximum when the sunlight is vertical to the surface and that is only at one point at any given time.

The computer models assume the Earth is a flat sphere with the energy distributed evenly across the surface. The total amount of energy received is the same for the sphere and the disc but the distribution is very different. On the sphere (Earth) much more is received at the Equator, than at the poles—a difference that creates heat at the Equator and cold at the Poles. This is a relatively simple model, but in reality it is enormously complicated by a variety of motions, not least the rotation and tilt, which varies the way the Earth is exposed to sunlight through the year. This change in tilt is given in most textbooks as 23.5°, but few people know it is constantly changing, thus causing climate change.

When I wrote about Uniformitarianism as the prevailing philosophical view of western science I had this changing tilt in my mind because it relates to climate change. Most are unaware that the orbit of the Earth around the Sun is not a fixed elliptical orbit or that the tilt changes. Without going into the details of these changes, it is important to know that science has known about these changes for over 100 years, but they are still not in most textbooks or in the public's understanding. It wasn't until the late 1980s that these changes, known collectively as the Milankovitch Effect, were generally accepted by the science.

Consider the changes in climate these changes make and how they complicate construction of a model. The IPCC do not include these effects in their computer models because they claim the time-scale is not appropriate, but if you are projecting for 100 years then they are significant relative to the possible fractional impact of human CO_2.

Understanding of the dynamics of the atmosphere is even more recent and incomplete. It leaps almost 2000 years from Aristotle—who knew there were three distinct climate zones, and about changing seasonal wind patterns—to the 18th century. In 1735 George Hadley used the wind patterns, recorded by English sailing ships, to create the first upper level diagram of circulation (Figure 7).

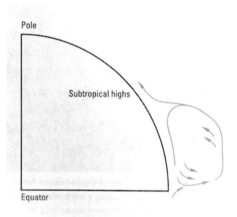

Figure 7

Hadley Cells

Restricted only to the tropics, these zones became known as Hadley Cells. Sadly, we know little more now than Hadley did.

The Intergovernmental Panel on Climate Change (IPCC) illustrates the point in Chapter 8 of the 2007 Report:

> *The spatial resolution of the coupled ocean-atmosphere models used in the IPCC assessment is generally not high enough to resolve tropical cyclones, and especially to simulate their intensity.*

The problem for climate science and modelers is that the Earth rotates. Its rotation around the sun creates the seasons, but the rotation around the axis creates even greater complexity. Without axis rotation, a simple single cell system (see Figure 8) with heated air rising at the Equator and moving to the Poles, then sinking and returning to the Equator, breaks down.

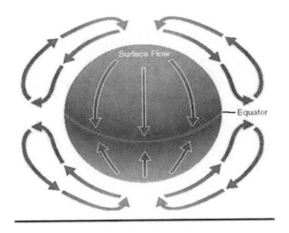

Figure 8

A Simple Single Cell[113]

[113] http://www.free-online-private-pilot-ground-school.com/Aviation-Weather-Principles.html

In the 1850s, William Ferrell attempted to improve understanding and proposed a three-cell system that still appears in most textbooks. This model shown in Figure 9 was convenient for teaching, but didn't work when research like tracking nuclear fallout from atmospheric explosions began in the 1960s.

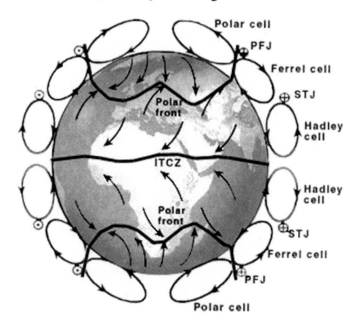

Figure 9

The Ferrell Three Cell Model

Figure 9 is inaccurate for a variety of reasons, but especially the difference in height of the cells. Figure 10 is a slightly better representation. Few people know the Tropopause, the boundary between the troposphere and the stratosphere is twice as high at the Equator as at the Poles.

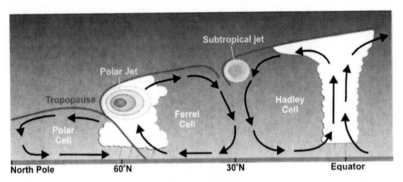

Figure 10

Varying Height of Tropopause in the Average Condition

The Tropopause height at the Poles varies between 7 km in winter and 10 km in summer—at the Equator the range is 17 to 18 km. The difference in seasonal range is because of the difference in seasonal temperature range, the summer winter difference is greater at the poles. How do you build even those simple dynamics into a computer model? Remember the heights vary with global temperate and that varies from year to year.

They created the Ferrell Cell to fill the gap, but the Ferrell Cell doesn't exist year round. Seasonally the cold air of the Polar Cell is denser and pushes warm air out of the way. In Figure 10, the boundary between Polar and Ferrell Cells is at 55°N, an average position. The range is from 35° N in winter to 65°N in summer.

Figure 11 shows a more recent attempt to approximate what is going on.[114]

---

[114] http://www-
das.uwyo.edu/~geerts/cwx/notes/chap01/tropo.html

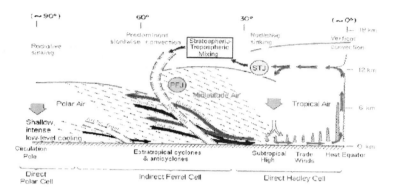

Figure 11

A Recent Model of a Cross-section through the Northern Hemisphere

Now it is called the *Indirect Ferrell Cell*, but the most important part is the discontinuity in the Tropopause and the *Stratospheric-Tropospheric Mixing*. This is important, because the IPCC doesn't include the critical connection between the stratosphere and a major mechanism in the upper troposphere in their models.

> Due to the computational cost associated with the requirement of a well-resolved stratosphere, the models employed for the current assessment do not generally include the QBO.

What are the computational costs associated with the requirement of a well-resolved stratosphere? This means they don't know what is going on, but that's true for most of the troposphere.

Climate models are mathematical constructs that divide the atmosphere into cubes as shown in Figure 5. It doesn't matter how many cubes you create for finer resolution because the data is simply not available, especially above the surface. The diagrams are cross-sections of average conditions, but they don't show the complex dynamics required for seasonal, annual, decadal and millennial changes and you realize the computer models are incapable of even approximating reality, but the problems don't end there.

Figure 12 shows a 3-dimensional view dividing the atmosphere into two air masses—the dome of cold air over the poles and in between the warm tropical air. The boundary is called the Polar front and marks the latitude of energy balance. In the cold dome more energy escapes than enters creating a deficit. In the warm air more energy enters then leaves creating a surplus. This difference between surplus and deficit drives the atmospheric system. The greatest difference occurs right at the Polar front and as a result the strongest winds are formed and indicated on the diagram as the Jet Stream.

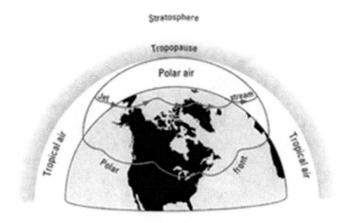

Figure 12

A General Diagram of the Relationship between Polar and Tropical, Air, the Jet Stream and the Polar Front

Figure13 indicates the pattern of the Jet Stream waves named after the discovery Carl Rossby. These Rossby Waves, shown in Figure 13, change in pattern between zonal or meridional flow with each creating very different weather patterns. They explain shifts in the variability of weather that has occurred since 2000 as the world cooled. IPCC models failed to understand this, but that is not surprising because it wasn't until 2007 that NASA admitted that the major cause of different ice conditions in the Arctic were due to changing wind patterns. It is just one of a multitude of limitations of their understanding of weather and climate.

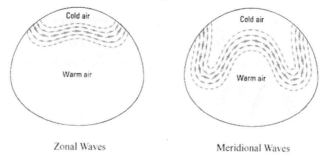

Zonal Waves   Meridional Waves

Figure 13

Comparison of Two Rossby Wave Patterns

Despite this the IPCC are certain about what has and will happen based on computer models that claim to replicate the atmosphere. This is a serious and unjustifiable claim.

Weather and Climate Forecast Failures

The most damaging evidence of the continued inadequacy of the models was outlined in a conference held in May 2008 at the University of Reading England.

Roger Harrabin, a BBC reporter with admitted sympathies to the IPCC work, wrote:

Chapter Seven

I have spent much of the last two decades of my journalistic life warning about the potential dangers of climate change...

...and reports they:

...plan a revolution in climate prediction.

What revolution? Belief that bigger and faster computers would solve the problems only illustrates their lack of understanding.

Julia Slingo of the University of Reading, a major centre for computer climate modelling, who became Chief Scientist for the United Kingdom Meteorological Office (UKMO) said:

> *We've reached the end of the road of being able to improve models significantly so we can provide the sort of information that policymakers and business require.*

Wrong! We never even got on the road.

> *In terms of computing power, it's proving totally inadequate. With climate models we know how to make them much better to provide more information at the local level...we know how to do that, but we don't have the computing power to deliver it.*

This is also wrong. It doesn't matter how big or fast the computer is if you don't have accurate data or understand climate mechanisms. The cost of computers is a serious limitation to open and effective research. This effectively restricts the work to government agencies, or academics funded by the government, which allows control by bureaucrats with political agendas. It also means it is necessary to stress the planet saving potential of the research.

No wonder the UKMO held a conference on March 18th, 2013 to discuss their failed seasonal forecast. They argued that the weather was unusual. It wasn't. The comment only underscores how little they know and the inaccuracy of their models.

Seth Borenstein lumped together several issues including predictions about Obama's re-election to prove computer model prognostications work.[115] He extended it to climate, likely because the U.S. public wasn't seeing global warming or climate change as problems anymore.[116] Corruption in leaked Climatic Research Unit (CRU) emails coupled with failed predictions, confirmed suspicions the sky wasn't falling.[117] As a result neither global warming nor climate change were mentioned in the U.S. election. Mainstream media and the President have put climate back on the table to justify unnecessary and damaging economic, energy, and environmental legislation.

Obama's re-election, attributed by some to the tropical storm Sandy became an opportunity to rewrite the record and put climate change at the center of funding and political control.[118] Borenstein used Sandy related events as the basis for unjustified claims to reinstate the failed work of the IPCC. He claims it was accurately forecast by computer models proving all the models work, that it was unprecedented and therefore proof of human activity as predicted by the IPCC.

Borenstein writes:

---

[115] http://bigstory.ap.org/article/predicting-presidents-storms-and-life-computer

[116] http://www.people-press.org/2012/01/23/public-priorities-deficit-rising-terrorism-slipping/

[117]

http://blogs.telegraph.co.uk/news/jamesdelingpole/100017393/climategate-the-final-nail-in-the-coffin-of-anthropogenic-global-warming/

[118] http://en.wikipedia.org/wiki/Hurricane_Sandy

*For about 40 years, climate scientists have used computer
models to predict what global warming will look like with
dead-on accuracy, said climate computer modeler Andrew
Weaver of the University of Victoria in British Columbia.*

It's false, but a cleverly worded statement because he is reported
saying what the effects of global warming will look like, which is
different from an actual prediction of conditions. Besides, he
knows the IPCC do not make "predictions."

No, the IPCC have deliberately misled the world about the
nature and cause of climate change for a political agenda.

There are two devastating reasons why the model output
should not be used for anything, especially policy. First, a
standard and required test of a model is its ability to predict.
Earlier I gave this as the simple definition of science. This is done
in models by a process called validation. You design a model based
on a known period of climate and then have it recreate (predict)
another known period.

No IPCC model has been validated—they cannot recreate
past climate conditions, therefore, they cannot make valid
predictions or even reasonable scenarios. They claim the models
are validated but what they actually do is keep adding or adjusting
variables until the model recreates the situation. This has nothing
to do with the actual processes and is called "tweaking." The
classic example was the attempts to make their models 'explain'
the cooling from 1940 to 1980. They did it by adding sulphates,
ostensibly from human activities, to act as cooling agent in the
atmosphere. The problem was temperatures started to increase
after 1980 but sulphur levels didn't decrease. Second, the most
basic assumption about human-caused climate change is that an
increase in CO_2 will cause an increase in temperature. Every
record shows that temperature increases before CO_2. Despite
this, they program computer models so temperature increases if
CO_2 increases, but that doesn't deter Harrabin, who concludes by

saying:

> *Political leaders have now agreed that they cannot wait*
> *for the modelling uncertainties to be ironed out. They*
> *have said they are convinced that emissions should be cut;*
> *but they just cannot agree who should do it.*

This statement underlines the bizarre thinking. The models are the sole source of evidence that there is a problem. They don't work, but we are going to act anyway. It's a standard strategy of environmentalists known as the Precautionary Principle.

How much longer can the IPCC maintain the charade before enough leaders understand the deceptions and show some backbone and shut them down? It is incredible that the IPCC and its manipulation of climate science continue to drive world energy and economic policies. How many more people must starve and economies collapse before they stop this environmentalist-driven exploitation?

Not only does the Emperor have no clothes; there is no Emperor.

Disastrous Policies Justified by False Claim that Climate Model *Predictions* are Correct

IPCC projections are worse than guesswork, which have a chance of being correct. IPCC tried to mask their complete failure by calling them projections and producing a range, but even the "Low" projection is wrong.

Figure 14 shows their most recent projections against the actual temperature. You can also see what the IPCC did when the data no longer fit the hypothesis. Up to about 2000 it was global warming, then when CO_2 continued to increase but when temperatures stopped increasing it became Climate Change. A

Chapter Seven

2004 leaked CRU (IPCC) email from the Minns/Tyndall Centre on the UEA campus said:

> In my experience, global warming freezing is already a bit of a public relations problem with the media.

To which Swedish alarmist Bo Kjellen replied:

> I agree with Nick that climate change might be a better labeling than global warming.

Proper science would require they consider the null hypothesis that something other than human CO_2 is causing warming. Instead, they moved the goal posts so the IPCC political objective was not abandoned.

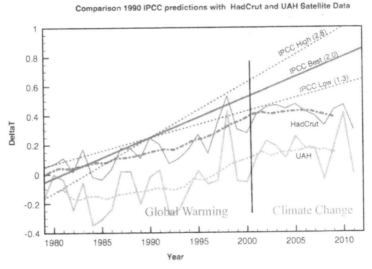

Figure 14

Comparing IPCC Predictions

IPCC climate models are the basis for claims of accurate predictions. They're inaccurate and the IPCC tell us why in Chapter 8 of the 2007 Physical Science Report.[119] Virtually nobody reads it. Instead they read the deliberately distorted Summary for Policy Maker (SPM) that gives completely unjustified certainty to model projections.

A more complete analysis of the distortions between the Science Report and the SPM shows the extent of the deliberate misdirection.[120] Most of the media were completely deceived as Borenstein's attempt at perpetuation proves. The climate predictions are wrong. Using them as the basis for draconian energy, economic and environmental policy brings disaster, as most European countries already know.

---

[119] http://www.ipcc.ch/publications_and_data/ar4/wg1/en/ch8.html
[120] http://drtimball.com/2012/climate-change-of-the-ipcc-is-daylight-robberyclimate-change-of-the-ipcc-is-daylight-robbery/

Chapter Seven

RAJENDRA PAUCHAURI

Part Two

What the IPCC Reports Say about the Computer Models

MOST DON'T UNDERSTAND models or the mathematics on which they are built; a fact promoters of human caused climate change exploited. They are used to 'prove' to the public that their science is unassailable. They are also the major part of the IPCC work not yet investigated by people who work outside climate science. Whenever outsiders investigate, as with statistics and the hockey stick, the gross and inappropriate misuses are exposed. The Wegman Report investigated the Hockey Stick fiasco but also concluded:

> *We believe that there has not been a serious investigation to model the underlying process structures nor to model the present instrumented temperature record with sophisticated process models.*

There are comments that provide insights. For example, in a comment about the multi-model comparison shown in Dr. Spencer's Figure 6, Ingvar Engelbrecht wrote:

> *I have been a programmer since 1968 and I am still*
> *working. I have been programming in many different*
> *areas including forecasting. If I have undestood [sic] this*
> *correctly this type of forecasting is architected so that*
> *forecastin [sic] day N is built on results obtained for day*
> *N—1. If that is the case I would say that its*
> *meaningless. It's hard enough to predict from a set of*
> *external inputs. If you include results from yesterday it*
> *will go wrong. Period!*

Of course, it is not necessary to know what is wrong within the
models. There failed predictions (projections) are clearly
displayed when compared to actual temperature.

More of the deception appeared in Assessment Report 5
(AR5) released in September of 2013. Despite a stretch of 17
years of slightly declining temperature, while CO_2 levels
continued to rise, they increased claims of certainty of their
results from 90+ percent to 95 percent. All sorts of excuses were
presented. One claimed the heat was hidden in the deep ocean.
Another said 17 years was an inadequate period of measure. The
point is the decline could not have happened at all according to
their science. Again the IPCC people knew because on October
12th, 2009 Trenberth said, "The fact is that we can't account for
the lack of warming at the moment and it is a travesty that we
can't."

Phil Jones, Director of the CRU and deeply involved with
what went on there and at the IPCC said:

> *We don't fully understand how to input things like*
> *changes in the oceans, and because we don't fully*
> *understand that you could say that natural variability is*
> *not working to suppress the warning. We don't know*
> *what natural variability is doing.*

The entire exercise of global warming and climate change is a deception. However, there are deceptions within the deceptions, with the most important one being that the IPCC produce essentially unassailable scientific predictions using computer models that represent reality.

There is hidden deception cleverly presented in clear view—a form of daylight robbery. This was achieved primarily by the difference between the definitive statements of the Summary for Policymakers (SPM) and *The Physical Science Basis* Report of Working Group I (WGI). The comments about the limitations of the models in Chapter 8 are in stark contradiction to the claims in the SPM. They underscore the political nature of the deception. They knew very few would read or understand the science report and they could be easily marginalized. With the limitations in the science report they could, if challenged, say they warned everybody.

MIT meteorology professor and former IPCC member Richard Lindzen summarized the process:

> *It uses summaries to misrepresent what scientists say; uses language that means different things to scientists and laymen; exploits public ignorance over quantitative matters; exploits what scientists can agree on while ignoring disagreements to support the global warming agenda; and exaggerates scientific accuracy and certainty and the authority of undistinguished scientists.*

In short, the IPCC clearly uses the Summary for Policymakers to misrepresent what is in the report.

When they compared the SPM to the AR4 Synthesis Report Armstrong and Green determined it:

> *...is a political document that downplays assessments of uncertainty from the scientific reports written by the main*

body of the IPCC, which themselves are far more subjective than the IPCC would have one believe. Equally important, both the IPCC's summaries and main reports omit much contrary evidence.[121]

It is only one of several deceptions, but unquestionably the worst because it was so effective.

Computer models produce IPCC projections as a central part of every Report. The process is progressive with each Report building on the last. They also, supposedly, introduce new findings. The difficulty is inclusions are selective because of the narrow definition of climate change and the rules that set a last date of publication. The last is reasonable because of meeting publishing schedules. However, there are instances of selective inclusions beyond the deadline or exclusion when the material was available in time.[122]

The IPCC produced the SPM for policymakers. Cleverly and cynically they disclosed the limitations of their work in the Physical Science Report of Working Group I. This means nobody can say they didn't acknowledge the limitations. One of the most serious is the lack of data and the way it is created. In the 2007 IPCC AR4, Chapter 8 Advances in Modelling they say:

Despite the many improvements, numerous issues remain. Many of the important processes that determine a model's response to changes in radiative forcing are not resolved by the model's grid. Instead, sub-grid scales are used to parametrize the unresolved processes, such as cloud formation and the mixing due to oceanic eddies. It

---

[121] Steven F. Hayward, Kenneth P. Green, and Joel Schwartz for the AEI.

[122] http://climateaudit.org/2008/05/25/wahl-and-ammann-2007-and-ipcc-deadlines/

continues to be the case that multi-model ensemble simulations generally provide more robust information than runs of any single model. Table 8.1[123] summarises the formulations of each of the AOGCMs used in this report.[124]

Parameterization is a fancy word for making up data when it doesn't exist. Because of the lack of data, this occurs for a majority of the surface grids and virtually all of the layers above the surface. This means they estimate average value for each grid and use those as the base for the model. Computer models produce estimates, which they use as 'real' data in another. Parameterization is probably applied to approximately 80 percent of the surface and at least 90 percent in the atmosphere.

Even with the grid scale used it is too large for major weather events such as thunderstorms.

The climate system includes a variety of physical processes, such as cloud processes, radiative processes and boundary-layer processes, which interact with each other on many temporal and spatial scales. Due to the limited resolutions of the models, many of these processes are not resolved adequately by the model grid and must therefore be parametrized. The differences between parameterizations are an important reason why climate model results differ.[125]

The problem is bigger than just cloud formation. For example,

---

[123] http://www.ipcc.ch/publications_and_data/ar4/wg1/en/ch8s8-2.html#table-8-1

[124] http://www.ipcc.ch/publications_and_data/ar4/wg1/en/ch8s8-2.html

[125] Ibid.

thunderstorms in the tropics are too small for the grids but they are a major mechanism of balancing and cooling the atmosphere. They distribute surplus heat from the Tropics to offset deficit heat energy in the Polar Regions (see Figure 11).

> *There is currently no consensus on the optimal way to divide computer resources among finer numerical grids, which allow for better simulations; greater numbers of ensemble members, which allow for better statistical estimates of uncertainty; and inclusion of a more complete set of processes (e.g., carbon feedbacks, atmospheric chemistry interactions).*[126]

They are saying there is inadequate computer capacity so large amounts of data are excluded and major mechanisms ignored.

Computer models came on the scene during my career. Gradually climate conferences were dominated by modelers and most did it arrogantly. They battled amongst themselves, with the person with the largest and fastest computer, dominating. Very few of them knew anything about climate but were apparently attracted by the challenge of modeling the complexity that is global climate. Some mathematicians and statisticians test theories using very long records of natural phenomena. However as A. N. Whitehead said:

> *There is no more common error than to assume that because prolonged and accurate mathematical calculations have been made, the application of the result*

---

[126] http://www.ipcc.ch/publications_and_data/ar4/wg1/en/ch8s8-2.html

to some fact of nature is absolutely certain.[127]

The modelers still dominate climate science through the IPCC, but do not heed Whitehead's warning. Not surprising considering their political motive.

I witnessed a good example early at a conference in Edmonton on Prairie Climate predictions and the implications for agriculture. Climate modeler Michael Schlesinger dominated as the keynote speaker. His presentation compared five major global models and their results. He claimed that because they all showed warming they were valid. Of course they did because they were programmed to that general result. The problem is they varied enormously over vast regions. For example, one showed North America cooling, while another showed warming. The audience was looking for information adequate for planning and became agitated, especially in the question period. It peaked when someone asked about the accuracy of his warmer and drier prediction for Alberta. The answer was 50%. The person replied that is useless, my Minister needs 95%. The shouting intensified.

Eventually a man threw his shoe on the stage. When the room went silent he said, "I didn't have a towel." We learned he had a voice box and the shoe was the only way he could get attention. He asked permission to go on stage where he explained his qualifications and put a formula on the blackboard. He asked Schlesinger if this was the formula he used as the basis for his model of the atmosphere. Schlesinger said yes. The man then proceeded to eliminate variables asking Schlesinger if they were omitted in his work. After a few eliminations he said one was probably enough, but you have no formula left and you certainly don't have a model. It has been that way ever since with the

---

[127]

http://www.mathacademy.com/pr/quotes/index.asp?ACTION=TOP &VAL=nature

computer models.

The IPCC only include those that create the results they want, namely proof of human causes of climate change. Also, every time you run the model it provides a different result because the atmosphere is chaotic. They resolve this by doing several runs and then using an average of the outputs.

As noted earlier, by leaving out very important components of the climate system they guarantee a human signal will result. As William Kinninmonth, meteorologist and former head of Australia's National Climate Centre describes them:

> ...current climate modeling is essentially to answer one question: how will increased atmospheric concentrations of CO_2 (generated from human activity) change earth's temperature and other climatological statistics? Neither cosmology nor vulcanology enter the equations. It should also be noted that observations related to sub-surface ocean circulation (oceanology), the prime source of internal variability, have only recently commenced on a consistent global scale. The bottom line is that IPCC's view of climate has been through a narrow prism. It is heroic to assume that such a view is sufficient basis on which to predict future 'climate'.

"Heroic" is an understatement. It is impossible; especially when the natural variable omitted is larger than the possible human variable. When you leave variables out of an equation, you no longer have an equation because one-side cannot equal the other. For example, the IPCC 2007 Report on their computer model of the atmosphere says, "Due to the computational cost associated with the requirement of a well-resolved stratosphere, the models employed for the current assessment do not generally include the QBO." This is the Quasi-Biennial Oscillation of upper level winds related to the pattern of El Nino.

The Myth of More Powerful Computers

Billions of taxpayer's dollars are wasted on politically motivated climate science, most of it on buying and running computers incapable of modeling global climate. They're incorrectly programmed so a CO_2 increase causes a temperature increase to fulfill the saying Garbage In Garbage Out (GIGO). In the real world temperature increases before CO_2, but the programmers need a political result. Naturally the temperature forecasts are consistently wrong, but that doesn't matter. They claim they're getting better and all they need are bigger, faster computers. It won't and can't make any difference, but they continue to waste money.

Recently Cray computers[128] produced the Gaea supercomputer for climate research at the National Oceanic Atmospheric Administration (NOAA). More commonly spelt Gaia, after the Greek Earth goddess of the religion of Environmentalism. They reinforced the image with a landscape panorama (see Figure 15).

---

[128] http://blogs.knoxnews.com/munger/2011/12/noaas-petascale-computer-for-c.html

Figure 15

Cray, "Gaea" Computer with Landscape Panorama[129]

It puts a false face on the waste of money. It will produce meaningless results like all the computers despite being one of the biggest and fastest. It has a 1.1 petaflops capacity. FLOPS means Floating-Point Operations per Second and peta is 10^{16} or a thousand million floating-point operations per second.

This sounds impressive, but is totally inadequate. At the Reading conference Shukla reported:

> *The current generation high-end computers for climate research have a capability of about 50 teraflops, which makes it possible to integrate a typical climate model with about 100 km horizontal resolution for 20 years in one day.*[130]

---

[129] Original Photo ORNL photos by Jay Nave
http://blogs.knoxnews.com/munger/2011/12/noaas-petascale-computer-for-c.html

130

http://www.ecmwf.int/newsevents/meetings/workshops/2008/ModellingSummit/presentations/Abstracts.pdf

So at 50 trillion floating point calculations per second they only study 20 years of record per day. Worse, each run using identical input yields different results so they average several runs. This is with a grossly simplified model with a grid so large that each covers very different climate regions. Shukla challenges:

> We must be able to run climate models at the same resolution as weather prediction models, which may have horizontal resolutions of 3- 5 km within the next 5 years. This will require computers with peak capability of about 100 petaflops.

It makes no difference: weather prediction models don't work either. Proponents argued weather predictions are different than climate predictions. They're not because climate is the average weather; wrong weather yields wrong climate.

Keynote speakers at climate conferences they dominated with the new technology. They fought among themselves for dominance not based on results because predictions were consistently wrong, but about who had the biggest and fastest computer. The challenge was to get funding.

Maurice Strong set up the IPCC through the World Meteorological Organization (WMO), which provided access to national funding for expensive machinery. It also meant appointment of climate modelers about whom Donna Laframboise commented:

> ...often called scientists, their work has little in common with traditional science.

Worse:

> Nor has the IPCC subjected climate models to rigorous evaluation by neutral, disinterested parties. Instead it

recruits the same people who work with these models on a daily basis to write the section of the Climate Bible that passes judgment on them. This is like asking parents to rate their own children's attractiveness.

The relationship between one country's climate modelers and the IPCC illustrates this point. George Boer is considered the architect of Canada's climate modeling efforts. As an employee of Environment Canada, he has spent much of his career attempting to convince the powers-that-be that climate models are a legitimate use of public money.

They are not. Canada's Auditor General identifies $6.36 billion "climate change funding announcements between 1997 and 2005" [131], but at what price? A December 13th, 2011 story provides an answer.[132] Environment and Sustainable Development Commissioner Scott Vaughan reports:

Environment Canada has failed to implement a strategic plan to improve its internal scientific research in areas ranging from managing air and water pollution to toxic chemicals.

Billions are spent on useless computers and climate change while not dealing with real problems. They're not alone, it's happening in national weather agencies round the world.

Despite all that money spent by Environment Canada, they have the worst performing computer climate models. Here is a plot (see Figure 16) comparing Canadian results with the IPCC ensemble average and the actual temperature.

---

[131] http://www.oag-ovg.gc.ca/internet/English/att_c20060901xe02_e_14568.html
[132] http://www2.canada.com/topics/news/story.html?id=5852120

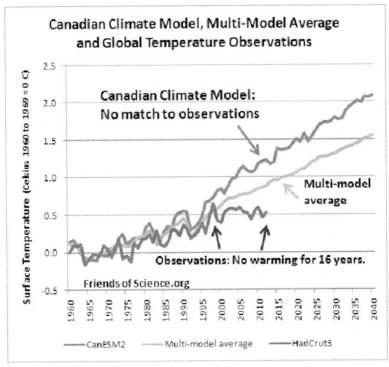

Figure 16

Canadian Climate Model

Climate change provided a vehicle for computer modelers to play their games. Here is what modeler Shukla wrote:

> *It is imperative that the summit recommends a realistic roadmap to enable and accelerate progress in climate modeling and prediction and to provide substantial and sustained support for enhanced workforce and computing resources. This is our moment! The problem of climate change has riveted the attention of the*

peoples of the Earth.[133]

Shukla's rationale is based on a false assumption:

> *Now we have before us, thanks to IPCC, a third discovery: humans are affecting the Earth's climate.*

This 2008 meeting concluded:

> *...the scientists had agreed that massive investment in computer and research resources was critical for revolutionising modeling capabilities so that predictions could capture the detail required to inform policy.*[134]

It was too late. The IPCC already informed policy in all their Reports with the 2007 Report claiming 90+ percent certainty human CO_2 was causing climate change. By the time the 2013 Report was due the hypothesis suffered from Huxley's syndrome—"*The great tragedy of science—the slaying of a beautiful hypothesis by an ugly fact.*" The fact was global temperatures declined from 1998 through 2013 while CO_2 continued to increase. Instead of admitting their science was wrong, they doubled down by claiming their certainty had increased to 95 percent.

There are two devastating reasons why the model output should not be used. First, a standard and required test of a model is its ability to predict. (In an earlier Part I gave this as the simple definition of science.) This is done in models by a process called

---

[133]

http://www.ecmwf.int/newsevents/meetings/workshops/2008/ModellingSummit/presentations/Abstracts.pdf

[134] http://buythetruth.wordpress.com/2009/06/24/met-office-fraudcast/

validation. You design a model based on a known period of climate and then have it recreate (predict) another known period. None of the IPCC models have ever been validated. They cannot recreate past climate conditions therefore they cannot make valid predictions or even reasonable scenarios. Second, the most basic assumption about human caused climate change is that an increase in CO_2 will cause an increase in temperature. Every record shows that temperature increases before CO_2. Despite this, the computer models are still programmed to have temperature increase if CO_2 increases.

The difference between what the IPCC acknowledge is missing or inadequate and the certainty of their conclusions is so stark it suggests it's clearly premeditated. It is beyond the scope of this book to look at all the limitations of the models, but a few will illustrate the problem. I will identify those that the IPCC manipulated to cover inadequacies or create a result.

There are major limitations of data like temperature, as Watts and D'Aleo explain in their detailed study of global records.[135] Knowledge of precipitation is even worse. When writing about precipitation data Bob Tisdale says, "Amazingly, there are few to no agreements among the three datasets when looking at the global data."[136]

In 2006 an attempt was made to forecast summer monsoon rains for the Sahel in Africa. They failed and concluded:

Climate scientists cannot say what has delayed the monsoon this year or whether the delay is part of a larger trend. Nor do they fully understand the mechanisms that

---

[135]http://scienceandpublicpolicy.org/originals/policy_driven_deceptio n.html

[136] http://wattsupwiththat.com/2013/07/23/no-consensus-among-three-global-precipitation-datasets/

govern rainfall over the Sahel. [137]

Worse, when they projected long-term trends one model said wetter, the other drier. Why?

> *One obvious problem is a lack of data. Africa's network of 1152 weather watch stations, which provide real-time data and supply international climate archives, is just one-eighth the minimum density recommended by the World Meteorological Organization (WMO). Furthermore, the stations that do exist often fail to report.*

They also concluded that the resolution of the models was inadequate and unaware of the underlying mechanisms.

However, there are many lesser variables critical to understanding what creates and determines the weather. Analogies are useful, but dangerous: witness the analogy of the atmosphere to a greenhouse. One limitation is the role of moisture in transferring heat and energy not considered in the greenhouse yet critical in the atmosphere.

Consider the role of moisture in controlling human body temperature. Few people know the skin is the largest organ of the body, performing an important function as the interface between the inner body and the atmosphere and controlling body temperature. The amount of moisture used in sweating is a fraction of the total amount in the body, but critical to controlling temperature.

The land and water surface of the Earth are the interface or skin between the subsurface and the atmosphere. The Earth is heated by shortwave energy from the sun that is stored in the surface. Energy for sweating is taken from the body, which cools the body. Evaporation of water from the surface takes heat stored

---

[137] "Waiting for the Monsoon" *Science* VOL 313, 4 Aug 2006.)

in the earth or water and transfers it to the atmosphere. The amount of soil moisture is estimated as "0.001% of the total water found on Earth," but like the sweat, it is critical to controlling temperature.[138]

The argument is belied by comments about missing variables such as this one from NASA that confirms the importance of soil moisture:

> *Soil moisture is a key variable in controlling the exchange of water and heat energy between the land surface and the atmosphere through evaporation and plant transpiration. As a result, soil moisture plays an important role in the development of weather patterns and the production of precipitation.*[139]

In 1992 the National Research Council suggested why weather forecast models don't work. Despite the importance of soil moisture information, widespread and/or continuous measurement of soil moisture is all but nonexistent.

> *The lack of a convincing approach of global measurement of soil moisture is a serious problem.*

This was in 1992 but little had changed when Chapter 8 of the 2007 IPCC Report appeared.

> *Since the TAR, there have been few assessments of the capacity of climate models to simulate observed soil moisture. Despite the tremendous effort to collect and homogenize soil moisture measurements at global scales (Robock et al., 2000), discrepancies between large-scale*

---

[138]http://www.spaceref.com/news/viewpr.html?pid=37477
[139]http://www.ghcc.msfc.nasa.gov/landprocess/lp_home.html

estimates of observed soil moisture remain.

Global climate models are composites of individual models for each component of the atmosphere and assume the smaller model output is real data and they know how it interacts with all other model inputs. Even that is a problem as Koster et al. explain:

> *The soil moisture state simulated by a land surface model is a highly model-dependent quantity, meaning that the direct transfer of one model's soil moisture into another can lead to a fundamental, and potentially detrimental, inconsistency.*

The NASA statement identifies another limitation of the IPCC model when they refer to "the exchange of water and heat energy between the land surface and the atmosphere."

Again the IPCC model fails as Chapter 8 notes:

> *Unfortunately, the total surface heat and water fluxes (see Supplementary Material, Figure S8.14) are not well observed.*

This means they don't have the data, but they also admit they cannot simulate the mechanisms involved.

> *For models to simulate accurately the seasonally varying pattern of precipitation, they must correctly simulate a number of processes (e.g., evapotranspiration, condensation, transport) that are difficult to evaluate at a global scale.*

Limitations of soil moisture data and mechanisms in the computer models invalidate any output they produce, but it's only one of many. What is outrageous is these gross inadequacies do not stop them claiming that:

> *Most of the observed increase in globally averaged*
> *temperatures since the mid-20th century is very likely due*
> *to the observed increase in anthropogenic greenhouse gas*
> *concentrations.*

"Most" and "very likely" are greater than 90 percent by their definition. These are high levels of certainty even in research based on solid data with reasonable understanding of the mechanisms. They're totally unjustified from the computer model inputs and outputs and the failure of every single prediction or scenario. Magnitude of the disparity suggests those who produce it are either scientifically incompetent or have created the result.

A large portion of the solar heat at the heat Equator is used for evaporation, changing the water from liquid to gas (water vapor). The heat used isn't lost but stored as latent heat and transported on the wind systems. Transfer of energy between the surface and the atmosphere, known as flux, is a major problem in the IPCC models. Their 2007 Report notes:

> *Unfortunately, the total surface heat and water fluxes are*
> *not well observed.*

Translation: they don't know how much heat and water moves in and out of the Earth's surface. They acknowledge it creates another problem.

> *These errors in oceanic heat uptake will also have a large*
> *impact on the reliability of the sea level rise projections.*

But threats of sea level rise are a major part of the fear and exploitation of the IPCC and their fellow Nobel winner Al Gore.

As the air rises it cools and condenses. The water vapor

converts back to liquid and the latent heat is released into the atmosphere. In the tropics this creates the major cloud form of cumulonimbus (thunderstorms), massive towering structures with powerful internal winds carrying vast amounts of energy through the atmosphere toward the Poles.

Global climate models divide the world surface into large rectangles. Essex and McKitrick prosaically note:

> *Not only can we not handle today's thunderstorms, but no such storm ever shows up, even in our very best computer climate models. Thus thunderstorms certainly are not dealt with from first principles in climate models either.*

The difficulty is:

> *...at every moment, there are thousands of active thunderstorms in the hot, moist places of the planet. There are tens of millions of them in any year. It should be clear that this great and constant roar of atmospheric air conditioning is an important part of the global energy budget should figure significantly into any model of the global climate however the mighty creature overhead, along with all his cousins, is too small to show up in even the biggest and grandest global climate models.*

Essex and McKitrick comment:

> *People who do serious climate calculations understand this problem and the fundamental scientific dilemma it implies. The only way to produce non-absurd calculations is to make up some ad hoc rules that insert or take away the energy, moisture or momentum has needed to produce sensible behavior.*

> *Even so, these made-up rules are not foolishly
> done. From the collective effects of sub-grid scale
> phenomena, parameterizations, empirical rules that
> mimic the overall effect of these phenomena fairly closely-
> are introduced.*

But they can't be close because the basic data is lacking and
mechanisms inadequately understood. As the IPCC report note
each model is parameterized differently so:

> *The differences between parameterizations are an
> important reason why climate model results differ.*

Inadequacies of modeling the Hadley Cell, a major mechanism in
producing global weather and climate, are enough to invalidate
the models and cause the failed predictions. The IPCC claim that:

> *Most of the observed increase in global average
> temperatures since the mid-20th century is very likely due
> to the observed increase in anthropogenic GHG
> concentrations.*

In IPCC jargon "very likely" means more than 90 percent certain,
but inadequate modeling of the Hadley Cell alone makes that an
utterly false claim. When the models projections failed, as they
did spectacularly in the 2013 Report, they only increased their
certainty.

How much longer can the IPCC maintain the charade? How
long before the IPCC and its machinations are understood by
enough leaders to elicit some backbone? It is incredible that the
IPCC and their manipulation of climate science continue to drive
world energy and economic policies. How many more people
must starve and economies collapse before this most egregious
exploitation driven by environmentalists is stopped?

CHAPTER EIGHT

Temperature Data and Data Manipulation

*I have no data yet. It is a capital mistake to theorize
before one has data. Insensibly one begins to twist facts to
suit theories, instead of theories to suit facts.*
—Arthur Conan Doyle. (Sherlock Holmes)

HUBERT LAMB, WHO many, including me, consider the
father of modern climate studies, established the Climatic
Research Unit (CRU). In his autobiography *Through All the
Changing Scenes of Life* he said that he created the Climatic Research
Unit (CRU) because:

> ...*it was clear that the first and greatest need was to
> establish the facts of the past record of the natural
> climate in times before any side effects of human activities
> could well be important.*[140]

---

[140] *Through all the Changing Scenes of Life: A Meteorologists Tale*, Hubert
Lamb, Taverner Publications, 1997, page 203

The amount of data is still inadequate, which poses a problem far beyond Lamb's concern.

It is trite, but still appropriate to quote Walter Scott's admonition about what a tangled web we weave when first we practice to deceive. By pre-determining the outcome of their scientific investigation the IPCC were increasingly forced to ignore, counteract, misinterpret, misinform and eventually adjust the data. It became increasingly necessary to become more devious. I've already discussed how their actions completely contradicted the scientific method of trying to disprove a hypothesis. The search was for data and research that proved the hypothesis even if it had to be manufactured. To illustrate the point, consider that in every case when they adjusted temperature data it always made the past colder and the present warmer.

The hockey stick was research designed to prove that the 20th century was the warmest in history. In that case, they eliminated the well-established Medieval Warm Period (MWP), which was clearly warmer. They rewrote history to prove their hypothesis. It is not the only example, albeit the most egregious. A global program to lower temperatures in the early part of the instrumental record changed the slope of the temperature curve. This made temperatures at the end of the 20th century appear warmer and suggested a more dramatic increase.

Anyone who knows anything about global climate knows the Earth has generally warmed since the 1680s. Politics has made that period, which covers the Industrial Revolution, of climatic interest as people wanted to prove that human production of CO_2 by their industry was causing warming. Proponents ignore the natural warming since the nadir of the Little Ice Age in the 1680s. The issue isn't the warming, but the cause.

Manipulation of records occurs by a variety of means including:

- Filling in missing data incorrectly or inappropriately
- Selecting stations that provide the desired result
- Incorrectly adjusting for instruments changes used over time
- Failing to or incorrectly adjusting for artificial heating by the urban heat island effect (UHIE)

The latter is an increase in temperature at the weather station as urban development surrounds the station. Professor Richard Muller tried to settle the problems in what also turned out to be deceptive with the Berkeley Earth Surface Temperature (BEST) project. It said:

> Our aim is to resolve current criticism of the former temperature analyses, and to prepare an open record that will allow rapid response to further criticism or suggestions.

Their actions and results altogether belie this claim and point to a political motive. The entire handling of their work has been a disaster. It is not possible to say it was planned, but it thoroughly distorted the stated purpose and results of their work. The actions are almost too naive to believe they were accidental, especially considering the people involved in the process. Releasing reports to mainstream media before all studies and reports are complete is unconscionable from a scientific perspective. They replicate the deceptive practice of the Intergovernmental Panel on Climate Change (IPCC) of releasing the Summary for Policymakers (SPM) before the Scientific Basis Reports with which it differs considerably.

Like the IPCC, the BEST panel appears deliberately selected to achieve a result or at least ensure a bias. It begins with the leader Richard Muller, who historically supported the anthropogenic global warming (AGW) hypothesis. There is only one climatologist in the group, Judith Curry, who only recently shifted from a vigorous pro AGW position to a more central and conciliatory position indicating awareness of the political implications. Involvement in the BEST debacle and especially her admission that early release of some results, before the peer-reviewed articles and supporting documentation, in other words the IPCC approach, was at her suggestion is troubling. Ms. Curry's comments indicate a very peripheral involvement in the entire process. This appears to suggest her participation was for public relations and supported by her comment in the data portion of the work. "I have not had "hands-on" the data." Failure to include a skeptical climatologist appeared to confirm the political objective.

Climate is the average of the weather, so it is inherently statistical. I am not a statistician, but sought professional advice whenever required. I watched the discipline change from simple analysis of the average condition at a location or in a region to a growing interest in the change over time. My experience with climate statistics taught me that the greater and more detailed the statistical analysis applied the more it underscores the inadequacies of the original data. It increasingly becomes an exercise in squeezing something out of nothing. The instrumental record is so replete with limitations, errors and manipulations that it is not even a crude estimate of the pattern of weather and its changes over time. Failure of all predictions, forecasts, or scenarios from computer models built on that database, confirm the inadequacies.

Problems start with the assumption that the instrumental measures of global temperature can produce any meaningful results. They cannot! Coverage is totally inadequate in space and time to produce even a reasonable sample. The map (see Figure 17) shows the pattern of Global Historical Climate Network (GHCN) stations from the BEST Report. It distorts the real situation. Each dot represents a single station but in scale probably covers over 500 Sq. km. They also don't show the paucity of stations in Antarctica, most of the Arctic Basin, the deserts, the rain forests, the boreal forest, and the mountains. Of course, none of these equal the paucity over the oceans that cover 70 percent of the world. It's a bigger problem in the Southern Hemisphere, which is 80 percent water.

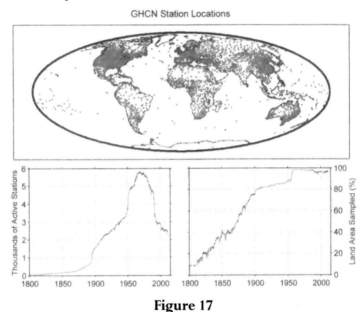

Figure 17

Areal coverage of GHCN stations[141]

BEST also shows the reduction in the number of stations after 1960. A major reason for the reduction was the assumption that satellites would provide a better record, but BEST didn't consider that record. It appears the main objective is to offset Ross McKitrick's evidence that much of the warming in the 1990s was due to a reduction in the number of stations (see Figure 18).

---

[141] Source: BEST Report, http://www.berkeleyearth.org/

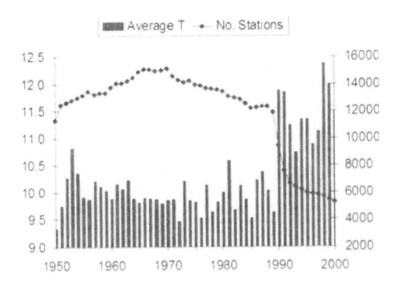

Figure 18

Global Temperature Plotted Against the Number of Stations

It is presented as a 'Surface' temperature record, but it isn't. It's the temperature in a Stevenson Screen (Figure 19) set according to the World Meteorological Organization (WMO) between 1.25 m (4 ft, 1 in) and 2 m (6 ft, 7 in) above ground. The difference is significant because temperatures in the lower few meters vary considerably as research shows. The 0.75 m difference means that you are not comparing the same temperatures.

Figure 19

Stevenson Screen[142]

Temperatures, sometimes to four decimal places, are used as if they are real, measured, numbers. Temperatures were recorded to half a degree because until thermocouple thermometers appeared any greater precision was impossible.

Most of the land data is concentrated in Western Europe and eastern North America so these latitudes dramatically over-represent the record. This is noteworthy because climate change is reflected most in these latitudes as the Circumpolar Vortex shifts between Zonal and Meridional flow and the amplitude of the Rossby Waves vary.

BEST used a subset of global temperatures, albeit a larger subset than anyone else. Because the full data set is inadequate, a bigger subset does not improve the analysis potential. Also, those who used smaller subsets did so to create a result to support a hypothesis. The BEST study apparently was designed to confirm the results and negate the criticisms.

---

[142] http://en.wikipedia.org/wiki/Stevenson_screen

Regardless of the BEST findings, the other 3 agencies achieved different results by the stations they chose, and the differences are significant. For example, one year there was a difference of 0.4°C between their global annual averages, which doesn't sound like much, but consider this against the claim that a 0.7°C increase in temperature over the last approximately 130 years. What people generally ignore is that in the IPCC estimate of global temperature increase produced by Phil Jones of 0.6°C, the error factor was ±0.2°C. That is a ±33 percent error, which makes the record and results meaningless because in many years the difference in global annual average temperature determined by different agencies is at least half the 0.7°C figure. In summation, all 4 groups selected subsets, but even if they had used the entire data set they could not have achieved meaningful or significant results.

The use of the phrase "raw temperature data" is misleading. What all groups mean by the phrase is the data provided to a central agency by individual nations. Under the auspices of the World Meteorological Organization (WM0), each nation is responsible for establishing and maintaining weather stations of different categories. The data these stations record is the true raw data. However, individual national agencies adjust the data before submission to the central record. They didn't use "all" stations or "all" data from each station. However, it appears there were some limitations of the data that they didn't consider, as the following quote indicates. Here is a comment in the preface to the Canadian climate normals 1951 to 1980 published by Environment Canada:

> No hourly data exists in the digital archive before 1953, the averages appearing in this volume have been derived from all available 'hourly' observations, at the selected hours, for the period 1953 to 1980, inclusive. The reader should note that many stations have fewer than the 28 years of record in the complete averaging.

BEST adjusted the data, but they are only as valid as the original data. For example, the National Institute of Water and Atmospheric Research (NIWA) produced the 'official' raw data for New Zealand and they 'adjusted' the "raw" data. The difference is shown in Figure 20. Which set did BEST use? Most nations have done similar adjustments.

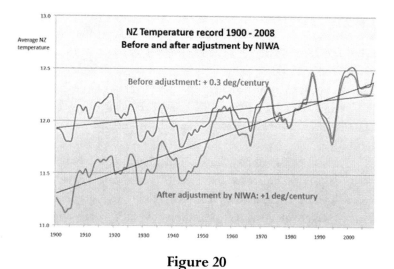

Figure 20

Temperature Record Adjustments for New Zealand

Chapter Eight

Here, in Figure 21, is the adjusted record for Maastricht:

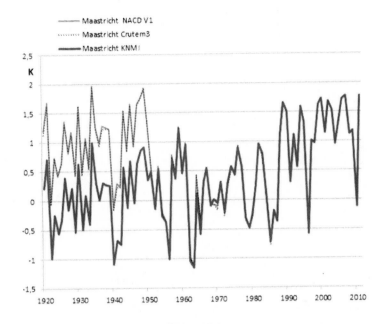

Figure 21

Temperature Record Adjustments for Maastricht

A plot of temperatures from the official source HadCRUT (Figure 22) shows the adjustments to the global temperature record. As the author notes:

> *I was curious about how HadCRUT4 differed from HadCRUT3. Look at the trend in the adjustment (black plot) starting around 2000. Also look at the adjustment to the first warm period (around 1918-1945) and to the cool period (around 1945-1977).*

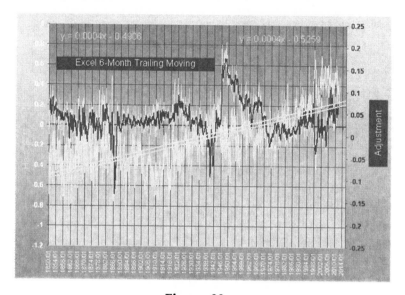

Figure 22

HadCRUT Temperature Adjustments[143]

They failed to explain how much temperature changes naturally or whether their results are within that range. The original purpose of thirty-year 'normals' was to put a statistically significant sample in a context. It appears they began with a mindset that created these problems and it has seriously tainted their work. For example, they say:

> *Berkeley Earth Surface Temperature aims to contribute to a clearer understanding of global warming based on a more extensive and rigorous analysis of available historical data.*

This terminology indicates prejudgment. Why global warming? It doesn't even accommodate the shift to "climate change" forced on

---

[143] Source: C. Bruce Richardson Jr.

proponents of anthropogenic global warming (AGW) as the facts didn't fit the theory. Why not just refer to temperature trends?

The project indicates a lack of knowledge or understanding of inadequacies of the data set in space or time or subsequent changes and adjustments. Lamb spoke to the problem when he established the Climatic Research Unit (CRU). BEST confirms Lamb's concerns. The failure to understand the complete inadequacy of the existing temperature record is troubling. It appears to confirm that there is incompetence or a political motive, or both.

Consider the case of the United States Historical Climate Network (USHCN) and its history of changing data identified by Steve Goddard. As he notes:

> USHCN2 uses a three step process to cool the past and warm the present. Going from the actual measured daily data to "raw monthly" reduces the decline. The TOBS (Time of Observation) adjustment flips the trend from cooling to warming, and then a final mysterious adjustment creates a strong warming trend.

On the website he uses a dynamic graph showing the adjustment. Here (see Figures 23 and 24) are two still frames of the raw and adjusted data.

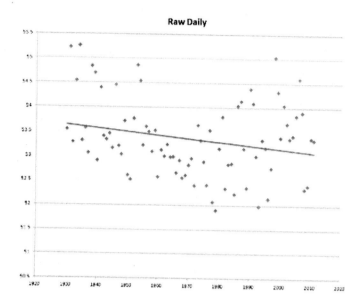

Figure 23

Raw Temperature Scatterchart

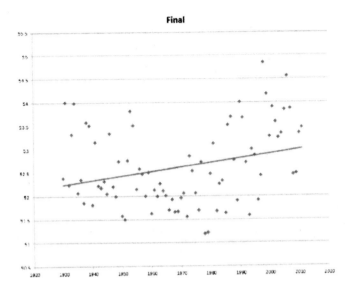

Figure 24

Final Temperature Scatterchart

Data Quality

I have already examined the decline in the number of weather stations and shown the impact on temperatures, but there are other problems. There is wholly inadequate spatial coverage, but the length of record is a bigger problem. The Goddard Institute for Space Studies (GISS) produced a graph (see Figure 25) as part of their definition of what was acceptable as a station for the official record.

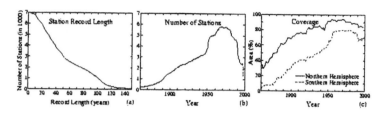

Figure 25

GISS Temperature Records

The left hand graph shows that there are fewer than 1,000 stations over 100 years in length. Almost all of these are in Western Europe and eastern North America, which indicates why Phil Jones temperature indicate are meaningless and why he had such a wide error factor.

People assume the best coverage, location and quality of stations was in the United States. Extensive research by meteorologist Anthony Watts[144] showed this was not the case. Figure 26 shows the results for 1007 of 1221 stations examined. Only 7.9 percent had an error factor less than 1°C. The remaining 92.1 percent had errors equal to or greater than 1°C.

---

[144] http://www.surfacestations.org/

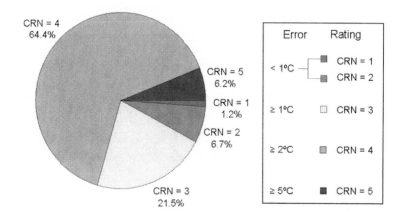

Figure 26

Station Error Rate

Two illustrations illustrate that the problem is contamination of the site by changes in the surroundings. The first station in Orland, CA (Figure 27) is well sited and unchanged.

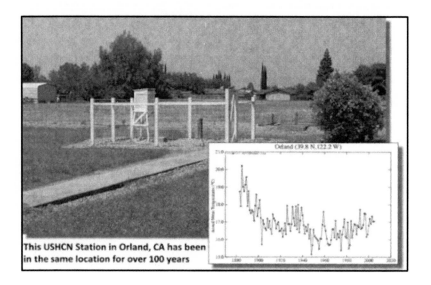

USHCN Station in Orland, CA

Figure 27

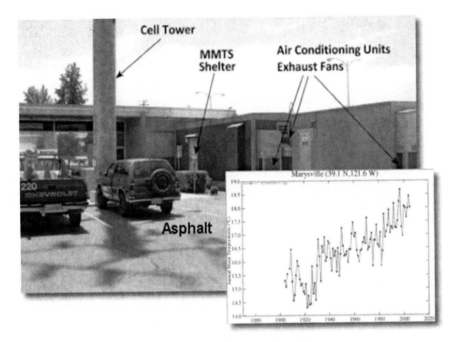

USHCN Station in Marysville, CA

Figure 28

The second station is in Marysville, CA (see Figure 28) and at the same site for the same length of time, but encroachment is evident in the picture and reflected in the temperature curve.

Anthony Watts joined with meteorologist Joseph d'Aleo to produce a comprehensive analysis of global temperatures, titled *Surface Temperature Records; Policy-Driven Deception?*[145] It is a devastating critique of the inadequacies of surface temperature record yet it is the basis of the entire IPCC studies and models.

---

[145]http://scienceandpublicpolicy.org/images/stories/papers/originals/surface_temp.pdf

Summary of the Distorted Temperature Record

1. Instrumental temperature data for the pre-satellite era (1850-1980) have been so widely, systematically, and unidirectionally tampered with that it cannot be credibly asserted there has been any significant "global warming" in the 20th century.

2. All terrestrial surface-temperature databases exhibit truly serious problems that render them useless for determining accurate long-term temperature trends.

3. All of the problems have skewed the data so as to substantially overstate observed warming both regionally and globally.

4. Global terrestrial temperature data are gravely compromised because more than three-quarters of the 6,000 stations that once existed are no longer reporting.

5. There has been a severe bias towards removing higher-altitude, higher-latitude, and rural stations, leading to a further serious overstatement of warming.

6. Contamination by urbanization, changes in land use, improper siting, and inadequately calibrated instrument upgrades further overstates warming.

7. Numerous peer-reviewed papers in recent years have shown the overstatement of observed longer term warming is 30-50% from heat-island contamination alone.

8. Cherry picking of observing sites combined with interpolation to vacant data grids may make heat-island bias greater than 50% of 20th-century warming.

9. In the oceans, data are missing and uncertainties are substantial. Comprehensive coverage has only been available since 2003, and shows no warming.

10. Satellite temperature monitoring has provided an

alternative to terrestrial stations in compiling the global lower-troposphere temperature record. Their findings are increasingly diverging from the station-based constructions in a manner consistent with evidence of a warm bias in the surface temperature record.

11. NOAA and NASA, along with CRU, were the driving forces behind the systematic hyping of 20th-century "global warming".

12. Changes altered the historical record to mask cyclical changes that could be readily explained by natural factors like multidecadal ocean and solar changes.

13. Global terrestrial data bases are seriously flawed and can no longer be trusted to assess climate trends or validate model forecasts.

14. An inclusive external assessment is essential of the surface temperature record of CRU, GISS and NCDC "chaired and paneled by mutually agreed to climate scientists who do not have a vested interest in the outcome of the evaluations."

15. Reliance on the global data by both the UNIPCC and the U.S. GCRP/CCSP also requires a full investigation and audit.

CHAPTER NINE

Exploiting the Rich and the Poor under the Guise of Saving the Planet

AT A 2004 conference of the Russian National Academy of Sciences, Sir David King, Chief Scientific Adviser to Tony Blair's government made the startling statement that *Global warming is worse than terrorism*. He was right, but not as he intended. The false premise promoted by the IPCC that human CO_2 caused global warming was used to terrorize and undermine developed nations in pursuit of Maurice Strong's goal of getting rid of them.

The former President of the Czech Republic, Vaclav Klaus, was the only world leader to understand the science and speak out about what Strong and his instrument the IPCC were doing. He was also immediately aware of communism and recognized what was happening. In a 2008 article for *The Australian*[146] he wrote:

> *I am afraid there are people who want to stop the economic growth, the rise in the standard of living*

---

[146]http://www.theaustralian.com.au/opinion/columnists/climate-alarmists-pose-real-threat-to-freedom/story-e6frg7ef-1111115772269

(though not their own) and the ability of man to use the expanding wealth, science and technology for solving the actual pressing problems of mankind, especially of the developing countries. This ambition goes very much against past human experience which has always been connected with a strong motivation to better human conditions. There is no reason to make the change just now, especially with arguments based on such incomplete and faulty science.

Rather than summarize again how Maurice Strong used the United Nations and specifically the IPCC to achieve his goal of getting rid of the industrialized nations, this chapter examines the devastation it has already brought. Reports of the IPCC, falsely presented as based on science, were used to scare the world, initially about global warming and later climate change. Politicians caught up with the need to appear green grasped at the output of the IPCC. They were thus vulnerable and easily fooled because they didn't understand and the entire objective of the IPCC was to mislead, misdirect and distort.

Instead of helping poor countries and poor people the machinations of Strong, Gore and the IPCC are reaping the rewards of their activities while the people pay the price. The people are paying in other ways as governments use IPCC reports to justify carbon taxes and other restrictive, punitive and expensive regulations. A vast industry has erupted as the UK newspaper the Telegraph reported:

Investing in climate change is proving to be profitable for governments, corporations, and investors from many sectors. Governments recent subsidies towards energy-efficient programs is bringing in newfound wealth for investors. In addition, the rising price of oil has been influential in pushing investments towards alternative energy sources. CEOs are taking charge in ways that were unforeseen.

So, the very people and industries the environmentalists and socialists despise are doing what they do best: make money.

Gore is different than Strong because his motive was initially personal and purely political. His family had money and status through his U.S. Senator father. Gore used—or rather misused—the misleading information of the IPCC Reports evidenced by them sharing the Nobel Prize. However, as his political ambitions receded he also began making a great deal of money through his involvement with carbon credit trading. Scientists of the IPCC may be involved in carbon trading, but they also benefit through a high profile, easier access to funding and easier promotion. All the relationships between Gore, Strong, and carbon credit trading[147] are well documented. In an article titled, *Al Gore's Carbon Crusade: The Money and the Connections Behind it,* the author notes:

> *Al Gore's campaign against global warming is shifting into high gear. Reporters and commentators follow his every move and bombard the public with notice of his activities and opinions. But while the mainstream media promotes his ideas about the state of planet earth, it is mostly silent about the dramatic impact his economic proposals would have on America. And journalists routinely ignore evidence that he may personally benefit from his programs. Would the romance fizzle if Gore's followers realized how much their man stands to gain?[148]*

There is nothing wrong with making money, however, promoting

---

[147] http://capitalresearch.org/pubs/pdf/v1185475433.pdf
[148] http://www.humanevents.com/2007/10/03/the-money-and-connections-behind-al-gores-carbon-crusade/

a demand through false information raises serious questions about ethics, morality and possibly even fraud. The problem is the IPCC and the few scientists who have controlled that agency provide him with some scientific justification for his position. It is clear neither Gore nor Strong understand the science, but that doesn't bother them, which is problematic.

Most of the problems evolve from the false claim that CO_2 is causing global warming/climate change. The claimed difficulties we now face evolve from foolish, ignorant attempts to resolve the non-existent problem. Everything was directed at reducing our use of fossil fuels particularly oil, natural gas and coal while promoting alternative fuels.

Global warming provided the perfect vehicle for environmentalists to spread their claim of human destruction of the planet. Previously they could only point at local or regional problems, but now they had a genuine "the sky is falling" cause that encompassed the entire globe. Now the demand was for global policies and Strong provided this at the Rio Conference in 1992 in the formation of the United Nations Framework Convention on Climate Change (UNFCCC or FCCC).

This agency created the Kyoto Protocol that became the battleground. Only the industrialized countries that Strong sought to eliminate had to reduce CO_2 emissions. They excluded Developing nations and arranged payments from the sinful industrialized nations as penance. It was the transfer of capitalist wealth the socialist Strong foresaw. Futility of the exercise was that if all nations participated and met their original targets no measurable difference in atmospheric CO_2 would occur yet that was the purported objective. Several nations saw the problems implementing Kyoto would create. The U.S. Senate voted 95-0 against ratification even though Al Gore was Vice President at the time. It reached a critical point when a failure of Russia to sign meant the Protocol would not be implemented.

The Protocol is now dead and there is no apparent successor.

However, this is not because of the false science. It is because the countries excluded from the Protocol, particularly India and China, have become the dreaded industrialized nations Strong opposes. Despite attempts by the few scientists who control the IPCC to push an alarmist Summary, China was a leading opponent and forced a softening of the final document so that even mainstream media noticed (see Part 8). The naiveté and political tunnel vision of Strong ignored the fact that every country in the world wanted to industrialize and emulate the United States. Or did he? In 2005, the Pittsburgh Tribune reported:

> *Recently, Strong was looking for an apartment in Beijing, where his Canadian interests are already enmeshed with the Chinese Red Army.*

A 2006 report said he formed a company with George Soros to import cheap Chinese made cars into the North American market. As the Tribune summarized:

> *Maurice Strong is the fox that was invited into the henhouse—and given the tools to redesign it for his own interests.*

Actually, he invited himself in and his "redesign", through the UN and the IPCC, did not stop global warming or climate change, but brought serious global problems. IPCC identification of CO_2 as the major culprit of environmental damage has:

- Allowed an unfounded and unwarranted attack on fossil fuels and exploitation of the false idea we are running out, especially of oil.
- Caused governments to promote alternate fuels as if they are the replacement solution when most are not viable

alternatives.

- Caused governments to provide massive direct or indirect subsidies that distort the value of these alternatives so that accurate cost benefit analysis is essentially impossible.
- Caused governments to provide subsidies for biofuels that seriously impeded world food production and are leading to starvation.
- Caused governments to identify CO_2 as a pollutant and seek its reduction when it is essential to plants and a reduction would put them in jeopardy.
- Caused many governments to restrict or ban development of most fossil fuel energy sources.
- Caused governments to spend billions on climate research to stop climate change when it is impossible.
- Caused a diversion of money to climate research better spent on real and identified pollution problems.
- Allowed environmentalists to bully whole societies into adopting inappropriate policies and ideas.
- Caused unnecessary increases in transportation costs that resulted in a higher cost of living that especially impacts the poor and middle class.
- Caused an increase in travel costs that were beginning to become affordable for most people.
- Caused extensive and unnecessary fear among people, but especially children.

Strong and the IPCC exploited fear with threats of impending doom due to global warming/climate change, but they also exploited the lack of knowledge about science and especially climate science. Governments, eager to be green, unknowingly introduced policies to reduce greenhouse gases that are undermining the developed nations as effectively as terrorism.

The title of Vaclav Klaus' book *Blue Planet in Green Shackles* succinctly summarizes the broader problem, which he summarizes as follows:

> *Future dangers will not come from the same source. The ideology will be different. Its essence will nevertheless be identical: the attractive, pathetic, at first sight noble idea that transcends the individual in the name of the common good, and the enormous self-confidence on the side of its proponents about their right to sacrifice the man and his freedom in order to make this idea reality. What I had in mind was, of course, environmentalism and its present strongest version, climate alarmism.*
>
> *There is no scientific justification for any energy or economic policies designed to reduce greenhouse gases or stop warming or climate change. CO_2 from human or natural sources is not causing global warming or climate change. Beliefs that it is are solely the product of the IPCC and their computer models, an agency and approach set up to mislead the world. Yes, as Sir David King said: global warming is worse than terrorism.*[149]

One of the qualifications on my resume now is "Environmentalist". Indeed, it is a title everyone can put after their name. We are all environmentalists to greater or lesser degrees. It is an outrage that certain people and groups have usurped this title and implied that only they care about the environment. While this book outlines the role of the Intergovernmental Panel on Climate Change (IPCC) in manipulating climate science, it did this within the cloak and dominance of environmentalism as the new paradigm in the

---

[149]http://news.bbc.co.uk/cbbcnews/hi/world/newsid_3382000/338
2019.stm

western view of the world.

The message the IPCC pushed suited the environmentalists. It enabled them to hide their activities from the usurped moral high ground of saving the planet with scientific "proof". They could isolate those who dared to question the science as anti-environment or paid by the oil companies who were the cause of the major problem of climate change. While this happened, politicians were convinced by the bureaucrats representing their country as members of the IPCC. Politicians were easily bullied because they didn't understand the science and wanted to be 'appear' green.

Primarily due to the actions of the IPCC, nobody tested the scientific theory that human CO_2 caused warming/climate change known as the Anthropogenic Global Warming (AGW). Rather, Maurice Strong set up the organization through the UN to perpetuate the unproven theory. They designed the IPCC mandate so they only looked at human causes of climate change, but the media and the public believe they are looking at natural climate change in total and scientifically. Rules guaranteed the message to the media; they created the illusion that they practiced and accepted only peer-reviewed science. Donna Laframboise and others exposed this falsehood.[150] They guaranteed the pre-eminence of the political message over the science by writing a rule to release the Summary for Policymakers before the Science Report. Another rule required the Science Report agree with the Summary. The final product achieved the result of deception in full daylight.

Starting in 1990, the IPCC produced Reports each increasing the probability that there was clear evidence of a human cause of initially global warming and then climate change.

Scientists discredited IPCC claims, but it didn't stop the

---

[150] http://wattsupwiththat.com/2010/01/24/the-scandal-deepens-ipcc-ar4-riddled-with-non-peer-reviewed-wwf-papers/

process. Release of the 2001 Report drove a deadly combination of events and misconceptions forward, overriding any attempts to point out the errors, omissions and deliberate misdirection. The media bought into the unproven theory and hysteria. Massive amounts of government funding drove research in a singular, but erroneous direction. Scientists who challenged were attacked and systematically marginalized. Nations developed environmental policies based on the misleading information of the IPCC. Al Gore's movie, which is based on the false information of the IPCC, fooled people on a global scale. Unnecessary and potentially destructive policies, including the Kyoto Protocol, were promoted as necessary to save the planet. Environmental and climate change hysteria took hold and reason was out the window.

The switch from global warming to climate change began about 2002, as natural events did not agree with what the computer models had predicted. However, the model predictions of future warming held sway while climate change allowed them to point to any event as proof. Warmer. Colder, Wetter, Drier, More Severe Weather, Less Severe Weather, it didn't matter now, humans caused it all. Now they had established the practice of claiming natural events as unnatural so they could never appear wrong. This drew public attention away from the failure of the model predictions, but the experts were continuing their focus.

For example, Dr. Donald Dubois noted:

> *If the major climate models that are having a major impact on public policy were documented and put in the public domain, other qualified professionals around the world would be interested in looking into the validity of these models. Right now, climate science is a black box that is highly questionable with unstated assumptions and model inputs.*

Chapter Nine

Or, as Dr. David Wojick explained:

> *The public is not well served by this constant drumbeat of false alarms fed by computer models manipulated by advocates.*

False assumptions guaranteed failure of the models of the climate system. Manipulation and deception included omissions. These exclusions increased the focus on CO_2, which was always the objective. The problem is CO_2 is not causing global warming or climate change at all. Politics has not caught up with the science primarily due to the machinations of the IPCC and a few political scientists so governments continue to deal with a non-existent issue.

What is wrong with the CO_2 argument?

AGW advocates and governments talk about reducing greenhouse gas, but they mean CO_2. Few know it is less than 4% of all the greenhouse gases and the human portion is just a fraction of the 4%. Indeed, the amount we produce is within the error factor of the estimates of three natural sources.

	Gt C p.a.
EMISSIONS:	
Humans and Animals	45 to 52
Oceans' outgassing (tropical areas)	90 to 100
Volcanic and other ground sources	0.5 to 2
Ground bacteria, rotting and decay	50 to 60*
Forest cutting, forest fires	1 to 3
Anthropogenic emissions (2010)	9.5
TOTAL	**196 to 226.5**
* other published values: 26 Gt, 86-110 Gt	
UPTAKE:	
Vegetation on land	55 to 62
Oceans' uptake	87 to 95*
Algae, phytoplankton, seaweed	48 to 57
Weathering (Silica minerals)	3 to 6
TOTAL	**193 to 220**
* other published values: 68-110, 30-50 Gt	

Source: Dr. Dietrich Koelle

Table 2

Table 2 shows the range of estimates of natural CO_2 and human production in 2005. Accuracy has not improved. Notice the human contribution is within the error range of two of the natural sources—Oceans and Ground bacteria. They tried to claim they could identify fossil fuel CO_2 but it is the same as for rotting vegetation. In other words if everyone left the planet but one scientist remained to measure the difference in atmospheric CO_2 they would not be able to measure any difference.

Many problems exist with the AGW theory, but there is one that destroys it altogether. The most fundamental assumption of the theory is that an increase in CO_2 will cause an increase in temperature. In fact, every record for any period and duration shows exactly the opposite happens—temperature increases before CO_2. This assumption, programmed into the computer models, continues to have a CO_2 increase cause a temperature increase. They dare not change it because it takes the focus away from CO_2, the sole target of the IPCC from its inception.

The crisis in confidence about the climate debate is good news—it is the certainty of people like Palmer and the IPCC that has created the crisis and will create the loss of credibility. The larger problem is the loss of credibility for science on all issues especially the environment.

The small group who have controlled the IPCC are unlikely to change their tune. Despite overwhelming real world evidence of cooling since 2001 while CO_2 levels continue to rise they push their agenda. Leo Tolstoy explained this phenomenon some 100 years ago when he wrote:

> *I know that most men, including those at ease with problems of the greatest complexity, can seldom accept even the simplest and most obvious truth if it be such as would oblige them to admit the falsity of conclusions which they delighted in explaining to colleagues, which they have proudly taught to others, and which they have woven, thread by thread, into the fabric of their lives.*

Or, as H L Mencken said:

> *The whole aim of practical politics is to keep the populace alarmed (and hence clamorous to be led to safety) by menacing it with an endless series of hobgoblins, all of them imaginary.*

Climate change will lead to a "fortress world" in which the rich lock themselves away in gated communities and the poor must fend for themselves in shattered environments, unless governments act quickly to curb greenhouse gas emissions, according to the vice-president of the Intergovernmental Panel on Climate Change (IPCC).

MICHAEL MANN

CHAPTER TEN

How the World Found out What Was Going on in IPCC Climate Science

I have known a vast quantity of nonsense talked about bad men not looking you in the face. Don't trust that conventional idea. Dishonesty will stare honesty out of countenance, any day in the week, if there is anything to be got by it.
—Charles Dickens

AS EXPLAINED IN Chapter Four, Maurice Strong set up two streams, the political with Agenda 21 and the scientific with the IPCC, to advance the ideas of The Club of Rome and its surrogate Agenda 21. These were continued, combined and implemented through UNFCCC and the Conference of the Parties (COP).[151]

---

[151] The COP is the link between the IPCC and policy as they explain: "The Conference of the Parties (COP) is the "supreme body" of the Convention, that is, its highest decision-making authority. It is an association of all the countries that are Parties to the Convention. The COP is responsible for keeping international efforts to address climate change on track."

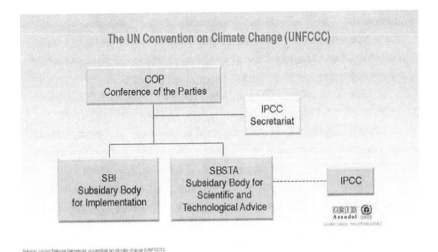

Figure 29

UN Convention on Climate Change

Whenever the COP meets they base their work on the results of the IPCC. The first COP for climate change was held in Berlin in 1995. Other important COP meetings include COP 3 that formalized the Kyoto Protocol, COP 7 that detailed rules for implementation and COP 13 in Bali that set goals for post 2012. This target was important because if Kyoto was not approved by that date it expired. Two meetings, COP 15 in Copenhagen and COP 17 in Durban, were critical. If the plans for Kyoto were going to be derailed it had to be prior to these meetings, especially Copenhagen. The meeting was scheduled for December 7-18th, 2009.

In November of 2009, somebody released 1,000 emails obtained from the Climatic Research Unit at the University of East Anglia. We still don't know who it was but the result was

http://unfccc.int/essential_background/convention/convention_bodie s/items/2629.php devastating and effectively undermined the objective of the Copenhagen meeting. Suddenly the COP, whose work is totally based on the findings of the IPCC, was confronted with evidence of corrupted science.

Like its namesake Watergate, the cover-up of Climategate, so named by British journalist James Delingpole, amplified the disgraceful behavior of the original disclosure. It began almost immediately. Phil Jones, Director of the CRU, called the police and said their files had been stolen. This was important because legally and in terms of public perception, hacked files are different than those exposed by a whistleblower revealing wrongdoing. Ironically, it helped him from that perspective but in saying they were stolen he had to confirm to the police that they were CRU files.

The University of East Anglia hired Neil Wallis of *Outside Organization* to handle the fallout from the emails leaked from the Climatic Research Unit (CRU) in November 2009.[152] Wallis, was a former editor at the *News of The World* arrested in connection with the phone hacking scandals that led to the resignation of London Metropolitan Police Commissioner and Deputy Commissioner as well as Andy Coulson, Prime Minister Cameron's press secretary. University spokesperson Trevor Davies said it was a "reputation management" problem, which he claimed they don't handle well. Why does a Scientific Centre like the CRU or a University need a PR organization? Apparently they didn't consider telling the truth.

There is a distinct boundary between those who understand the science and know what the emails say, and those who don't have the knowledge and claim they are of no consequence. If the latter also have a political bias, tunnel vision is narrowed. It is not surprising that those involved and their acolytes claimed they are

---

[152] http://climateaudit.org/2011/07/20/east-anglias-toxic-reputation-manager/

not significant. The trouble is Jones already acknowledged they were real. Later he attempted to downplay their significance claiming the emails were normal banter between scientists. Sir Muir Russell in his investigation report said the emails were "robust" and "typical of the debate that can go on in peer review." If he really believes this it may explain why the entire group became so corrupted—it supports the concept of Groupthink discussed later.

Who Released the Files? Knowledge and Access to the University Computer System

Whoever released the files had access to the UEA computer system and knew which files and emails were significant.[153] After detailed analysis Canadian network engineer Lance Levsen showed convincingly the source was someone within the university. He concluded:

> For the hacker to have collected all of this information s/he would have required extraordinary capabilities...to crack an Administrative file server to get to the emails and crack numerous workstations, desktops, and servers to get the documents.

Despite the Norfolk Police claim that they had eliminated any internal source, there are several internal candidates. The police would not know what questions to ask in the same way most people getting the leaked information had little idea of its significance. Chief internal candidate for me was Keith Briffa or possibly somebody working with him. Emails show his conflicts within the group.

On October 5th 2009, Wigley wrote to Jones:

---

[153] http://www.smalldeadanimals.com/FOIA_Leaked/

> *It is distressing to read that American Stinker item. But Keith does seem to have got himself into a mess. As I pointed out in emails, Yamal is insignificant...I presume they went thru papers to see if Yamal was cited, a pretty foolproof method if you ask me. Perhaps these things can be explained clearly and concisely—but I am not sure Keith is able to do this as he is too close to the issue and probably quite pissed of [sic]. I think Keith needs to be very, very careful in how he handles this. I'd be willing to check over anything he puts together.*

Jones forwarded the email to Briffa.

Briffa's dislike of Mann goes back a long way. On June 17th, 2002 Briffa wrote to Dr. Edward Cook about a letter involving Esper and Michael Mann:

> *I have just read this lettter—and I think it is crap. I am sick to death of Mann stating his reconstruction represents the tropical area just because it contains a few (poorly temperature representative) tropical series. He is just as capable of regressing these data again any other "target" series, such as the increasing trend of self-opinionated verbage (sic) he has produced over the last few years, and...(better say no more)*

Cook responds:

> *We both know the probable flaws in Mike's recon (reconstruction), particularly as it relates to the tropical stuff.... It is puzzling to me that a guy as bright as Mike would be so unwilling to evaluate his own work a bit more objectively.*

On September 22nd, 1999 Briffa again confronted Mann in a long email that included the comment, "I believe that the recent warmth was probably matched about 1000 years ago." Treasonous words—for Mann's hockey stick paper that claimed no medieval warm period existed. Mann appeared to back off. He wrote, "Walked into this hornet's nest this morning! Keith and Phil have both raised some very good points." In reality he puts Briffa down again:

> *SO(sic) I think we're in the position to say/resolve somewhat more than, frankly, than Keith does, about the temperature history of the past millennium. And the issues I've spelled out all have to be dealt with in the chapter.*

One cynical comment from Mann says, "And I certainly don't want to abuse my lead authorship by advocating my own work." It's an apparent example of Mann's behavior, because he did just that in the IPCC 2001 Science Report and Summary for Policy Makers. It was an action Professor Wegman admonished in his Report. His First Recommendation says:

> *It is especially the case that authors of policy related documents like the IPCC report, Climate Change 2001: The Scientific Basis, should not be the same people as those that constructed the academic papers.*[154]

Wigley didn't help. Here is the first part of a belittling email from Wigley to Briffa on January 10th, 2006:

> *Thanx for this. Interesting. However, I do not think your*

---

[154] *"Ad Hoc Committee Report on the 'Hockey Stick' Global Climate Reconstruction"* for the Chair of the US Congress Committee on Energy and Commerce and the Chair of the Subcommittee on Oversight and Investigations.

response is very good. Further, there are grammatical and text errors, and (shocking!!) you have spelled McKitrick wrong. This is a sure way to piss them off. Typical of Wigley's patronizing way of talking to wayward CRU members.

Conflict continued as Briffa expressed his concern. Mann made some overtures, but on April 29th 2007 Briffa responded:

I found myself questioning the whole process and being often frustrated at the formulaic way things had to be done—often wasting time and going down dead ends. I really thank you for taking the time to say these kind words. I tried hard to balance the needs of the science and the IPCC, which were not always the same.

What damning commentary about what the CRU and the IPCC were doing?

Briffa may have worked with the Information Officer at the University who was under pressure for Freedom of Information (FOI) requests. In September we learned Briffa was ill. Did this give him time to think about what was happening? Maybe, but his treatment by Mann and the sinking ship was likely an impetus. Whatever the correct answer, any reading of the emails show they were anything but normal correspondence between colleagues.

Timing of the release explains how and why they selected just 1,000 emails to quickly and incontrovertibly draw attention to the corrupt practices and science of the IPCC, even if they didn't understand the science. Then there was enough scientific malfeasance to underscore what some of us had known for a long time. In addition, most didn't, and still don't, know that the IPCC and CRU are essentially the same organization.

Because of Jones' actions the Norfolk police, a regional force, involved the national government through the National

Domestic Extremism Unit, surely another measure of the seriousness of what was involved in the files. This led to the University turning over all the files related to skeptics and their requests through Freedom of Information (FOI). Apparently the police and subsequent investigations accepted the CRU claims that requests for information were politically driven and caused hardship that diverted them from their work. When police interrogated skeptics, they asked about political affiliations.[155]

Why?

The idea of politics as the only motive developed because the CRU and the IPCC made global warming a purely political issue. Besides, why has motive got anything to do with the requests for scientific data and process, especially when funded by taxes and used to create potentially devastating policies?

Prior to leaking the emails to the world on November 19th, 2009 the person sent them to Paul Hudson, weather and climate change expert with the BBC and former UK Met Office employee. Hudson received them five weeks before on October 23rd, 2009.[156] Hudson never explained why he did not release them although he did confirm they were identical to the ones released later. Hudson knew the implications of the emails because he had written an article a month earlier titled, *Whatever happened to global warming?* It is likely this article, and his access to the world through his position with the BBC, explain his selection for receiving the leaks. It is also likely that previous admonitions about his views from CRU people prevented him from releasing the information. This article put a big target on his back and is typical of how members of the CRU dealt with transgressions and targeted transgressors.

---

[155] http://www.express.co.uk/posts/view/169578/College-gives-files-of-climate-change-sceptics-to-police

[156] http://www.dailymail.co.uk/news/article-1231763/BBC-weatherman-ignored-leaked-climate-row-emails.html

An October 11th 2009 email from Narasimha Rao to Stephen Schneider says:

> *You may be aware of this already. Paul Hudson, BBC's reporter on climate change, on Friday wrote that there's been no warming since 1998, and that pacific oscillations will force cooling for the next 20-30 years.*

Mann became aware and on the 12th wrote:

> *...extremely disappointing to see something like this appear on BBC. Its [sic] particularly odd, since climate is usually Richard Black's beat at BBC (and he does a great job). From what I can tell, this guy was formerly a weather person at the Met Office. We may do something about this on RealClimate, but meanwhile it might be appropriate for the Met Office to have a say about this, I might ask Richard Black what's up here?*

This is Mann at his bullying best. Hudson's failure was disappointing because he had credibility as a BBC weather presenter and former Met Office employee. As I will explain later, bullying and intimidation is a major vehicle in the entire environmental and climate science process of the last thirty years.

In an October 12th, 2009 leaked email, Kevin Trenberth responded to Hudson's question as follows:

> *Well I have my own article on where the heck is global warming? We are asking that here in Boulder where we have broken records the past two days for the coldest days on record. We had 4 inches of snow.*

Stephen Schneider, another active part of the CRU and IPCC group also communicated with Hudson as a BBC communiqué explains; as previously mentioned, Paul wrote a blog for the BBC website on October 9th entitled *Whatever Happened to Global Warming*[157] and There was a big reaction to the article—not just here but around the world. Among those who responded were Professor Michael E Mann and Stephen Schneider whose e-mails were among a small handful forwarded to Paul on October 12th. This seems like the explanation for Hudson's reticence then, but why later?

Later Hudson blogged that he only received some of the larger set released and could not confirm they were all genuine. However, after he failed to release the information the person who sent them released them through a Russian internet provider. The urgency was the impending meeting in Copenhagen. From there, the web page Air Vent picked up and released them. That triggered Jones' claim of a burglary, exposure of what skeptics had suspected for many years and justification for their FOI requests. What happened is they turned the story, likely with PR advice to blame those who requested the FOI.

There are many questions beyond the failure of the Norfolk police investigation. Does Hudson know who sent him the emails? Was he interrogated? Surely, it is easy to track his emails and determine the source. One can only assume that hiding the identity of the person who released the emails is a necessary part of the whitewash and cover-up. What happened to the concern that drove the 'leaker' in the first place? Was he convinced, as Hudson appears to have been, that silence is a wise choice?

A few emails suggest why Hudson went silent. He knew the wrath and reach of Michael Mann. As a CRU member noted on October 26th, 2003:

Anyway, there's going to be a lot of noise on this one,

---

[157] http://www.bbc.co.uk/blogs/paulhudson/2009/10/whatever-happened-to-global-wa.shtml

> *and knowing Mann's very thin skin I am afraid he will*
> *react strongly, unless he has learned (as I hope he has)*
> *from the past...*

He didn't, as his later reactions showed.

Why don't we know who released the emails from the Climatic Research Unit (CRU)? Why did the Norfolk Police working the national government through the National Domestic Extremism Unit fail to find out? What does it say about the vulnerability of their computer systems?

Is it an attempt by the CRU and the University of East Anglia (UEA) to divert even more attention from their involvement in this scandal? Is it part of the larger cover-up apparently orchestrated by the Royal Society? This raises the question of what the CRU had to hide. If there was nothing in the files of consequence then loss of the information had no currency. The British House of Commons' Science and Technology Committee perpetuated this idea by referring to the emails as "stolen" in their whitewash investigation of Jones' behavior. They didn't even take testimony from scientists qualified to address the problems with the science, yet still concluded the science was solid.

The answer is more likely that the whistleblower will disclose motive and chicanery well beyond what is disclosed in the actual emails. There is a distinct boundary between those who understand the science and know what the emails say, and those who don't have the knowledge and claim they are of no consequence. If the latter also have a political bias the tunnel vision is narrowed even more.

This led to the University of East Anglia turning over all the files related to skeptics and their requests through Freedom of Information (FOI). Apparently the police and subsequent investigations accepted the CRU claims that requests for information were politically driven and caused hardship that diverted them from their work, which sounds more like work of the spin-doctors.

The idea that politics was the motive only developed because the CRU and the IPCC had made global warming a purely political issue. Besides, what has motive got to do with requests for scientific data and process, especially when funded by taxes and used to create potentially devastating policies?

Professor Wegman recognized this problem as well in his recommendations.

> *Recommendation 1. Especially when massive amounts of public moneys in human lives are at stake, academic work should have a more intense level of scrutiny and review.*

> *Recommendation 2. We believe the federally funded research agencies should develop a more comprehensive and concise policy on disclosure.*

People at the Climatic Research Unit (CRU) in England brazenly defend the indefensible. They said there was no significance to the emails and they were normal banter between scientists. They used it to deceive the world and it worked with most people because they don't understand the science. Besides, as I introduced in the Preface, they had control over mainstream media particularly through Andrew Revkin and Seth Borenstein at the New York Times and latterly the Associated Press (AP), George Monbiot at the Guardian and Richard Black at the British Broadcasting Corporation (BBC).

Seth Borenstein, like Revkin at the New York Times, apparently has no journalistic integrity. Here is his email to the CRU gang on July 23rd, 2009:

> *Kevin, Gavin, Mike, It's Seth again. Attached is a paper in JGR today that Marc Morano is hyping wildly. It's in a legit journal. Watchya think?*

"Again" means there was previous communication. A journalist talking to scientists is legitimate, but like the emails tone and subjective comments, it is telling.

IPCC and CRU Are the Same Corrupt Organization with a Legacy of Billions of Dollars Wasted and Unmeasured Loss of Lives

Cost of the corruption of climate science by the IPCC is likely a trillion dollars already and there is no measure of lives lost because of unnecessary reactions like biofuels affecting food supplies. Stories appear about the corruption at the IPCC and others about the leaked emails from the Climatic Research Unit (CRU). Most people, including the media, don't realize the IPCC is the CRU. Some articles mention both but don't make the connection. Failure to make the connection allows people involved to develop defenses, withdraw from associations eliminate documents especially emails. Consider Jones' advice on requests for Freedom of Information (FOI):

> *If they ever hear there is a Freedom of Information Act now in the UK, I think I'll delete the file rather than send to anyone.*

> *We also have a data protection act, which I will hide behind. Tom Wigley has sent me a worried email when he heard about it—thought people could ask him for his model code. He has retired officially from UEA so he can hide behind that.*
> *IPCC is an international organization, so is above any national FOI. Even if UEA holds anything about IPCC, we are not obliged to pass it on.*
> *Can you delete any emails you may have had with Keith*

re AR4?

PS I'm getting hassled by a couple of people to release the CRU station temperature data.

Don't any of you three tell anybody that the UK has a Freedom of Information Act!

It is impossible to imagine defining these comments as normal banter among colleagues.

It Appears the Cover-Up and Intimidation Continue

On March 13th, 2013, someone with the pseudonym Mr. FOIA released the remaining 220,000 emails leaked from the Climatic Research Unit (CRU) at the University of East Anglia (UEA) to a select few. Since then we have heard nothing. Mr. FOIA suggested there was more useful information with examples, such as origin of the term "deniers" applied to those who questioned Intergovernmental Panel on Climate Change (IPCC) science. Why hasn't more information appeared? Has threat of legal action or other intimidation silenced further release and analysis of these emails?

It appears that the IPCC and its supporters have, again, effectively shut down revelation of how they directed and controlled climate science to a predetermined outcome. Release of 1,000 emails from the CRU in November 2009 revealed collusion, calumny and corruption of climate science. It effectively derailed the Copenhagen Conference of the Parties (COP 15) that planned to salvage the Kyoto Protocol. Typically, it triggered cover-ups that appear to continue. It is a pattern that results from believing Machiavelli or Saul Alinsky's dictum that the end justifies the means.

Release of 5,000 more emails in November 2010 effectively slowed further attempts at advancing a Kyoto like agenda at Cancun (COP 16) then Durban (COP 17). The remaining 220,000+ emails were released in March 2013 with over 1,000 comments accompanying the news on Wattsupwiththat. Most people anticipated greater clarification of what was done in manipulating climate science. To my knowledge there is silence to date. Why?

Some possible factors for the delay include:

- The emails were released to a select few and they apparently chose not to work on them.

- They apparently did not take Mr. FOIA's advice, "To get the remaining scientifically (or otherwise) relevant emails out, I ask you to pass this on to any motivated and responsible individuals who could volunteer some time to sift through the material for eventual release."

- The delay appears to support my contention that it was such a mammoth task, picking out the first 1,000 released with Climategate 1, it was only possible by an insider familiar with what went on. By insider I do not mean only someone working at CRU. It could be someone on the email listing with access to their computers.

- People with the emails are afraid of recriminations if they release more analysis.

- It is also likely that intimidation, a favorite tactic of those defending what has occurred, was employed. It takes several forms including creating terms to impugn the credibility. The term "denier" is an example. However, a more effective intimidation was the actual, or even threat, of legal action.

Chapter Ten

Shortly after the 220,000 emails were leaked Anthony Watts posted an update that may have inadvertently sent a chill through the community seeking the truth:

> *UPDATE8 3/19/13: Jeff Condon has received legal notice from UEA warning him not to release the password. So far, I have not seen any such notice. For those who demand it be released, take note.—Anthony*

It is reminiscent of Gavin Schmidt's comment in an email to Lucia Liljegren shortly after the first release in 2009:

> *Date: Thu, 19 Nov 2009 15:48:21 -0500*
> *From: Gavin Schmidt*
> *To: lucia liljegren*
> *Subject: a word to the wise*
> *Lucia, As I am certain you are aware, hacking into private emails is very illegal. If legitimate, your scoop was therefore almost certainly obtained illegally (since how would you get 1000 emails otherwise). I don't see any link on Jeff-id's site, and so I'm not sure where mosher got this from, but you and he might end up being questioned as part of any investigation that might end up happening. I don't think that bloggers are shielded under any press shield laws and so, if I were you, I would not post any content, nor allow anyone else to do so. Just my twopenny's worth.*

Is this friendly advice or a warning shot? The history of climate science seems to suggest the latter.

It is generally assumed that laws are designed to protect people and provide justice. They are a positive component of a civilized society, but the growing use of the law for intimidation is negative and troubling. In many jurisdictions this use of Strategic

Lawsuits Against Public Participation (SLAPP) are prohibited.

Even if a lawsuit is not filed, the intimidation factor is very effective. A lawyer's letter intimidates most people. They incorrectly think it is a legal document or they know it will involve unaffordable costs. Besides, who can afford to fight government or even corporations? More infuriating, governments use citizen's money to sue them. Citizens also effectively pay the corporations because their legal costs are tax deductible or they pass the cost on in their prices.

Tactics of intimidation are part of the claim that the science is settled. They are linked with the campaign to stifle debate. Al Gore prefaced his 2007 "settled" statement to the U.S. Congress with, "The debate is over." It must not be over on general principles because debate is essential to understanding and progress, as any honorable politician knows. It cannot be over in science because it is fundamental to the advance of knowledge and understanding. Surely it is easier to show what's wrong with the alternate view, but they don't do that instead they divert, deflect, attack individuals and intimidate. What is their fear? What do they have to hide? Is the truth inconvenient? It appears they believe and practiced Machiavelli's dictate.

A Surprisingly Small Cast

Unfortunately, universities and governments participated in whitewashing the behavior of prominent individuals like Phil Jones and Michael Mann among others. Nobody else involved with the scandal is facing even biased internal investigation. Many are not mentioned in the limited media reports on the scandal. People like Mike Hulme, Tom Wigley, Benjamin Santer, Kevin Trenberth, Keith Briffa, Malcolm Hughes, Raymond Bradley, John Holdren, Jonathan Overpeck, Caspar Amman, Michael Oppenheimer, Tom Crowley, Gavin Schmidt, William Connolley, Tim Osborn, Thomas Karl, Andrew Weaver, Eric

Steig and all names on the CRU emails require investigation. They had to know what was going on, partly because they all used the same vehicles of attack and deception. By investigating only two individuals, the collective culpability of the CRU and the IPCC goes unchallenged. Investigation of two individuals underscores the false claim there are one or two "bad apples" but the overall science is unaffected. The IPCC received a Nobel Prize collectively; they must bear the collective blame.

There are also those in government who acted in extremely questionable ways. Chief among these are members of the United Kingdom Meteorological Office (UKMO) including John Mitchell. He was review editor of the IPCC and initially denied access to information then claimed it was erased. The UKMO later said the information existed but said it was protected information. The Telegraph newspaper said:

> *Documents obtained by The Mail on Sunday reveal that the Met Office's stonewalling was part of a coordinated, legally questionable strategy by climate change academics linked with the IPCC to block access to outsiders.*

What was the role of government officials who selected their country's representatives to prevent skeptics participating? Such was apparently the case in Canada, the UK and likely the US. UK Science advisor John Beddington has already said failure to include skeptics was a mistake:

> *I don't think it's healthy to dismiss proper scepticism. Science grows and improves in the light of criticism. There is a fundamental uncertainty about climate change prediction that can't be changed.*

The problem is exacerbated when it is still an active policy of the government. The release, content, and actions of 'official' agencies, as well as the mainstream media to Assessment Report 5 (AR5) saw exactly the same behavior. Nothing has really changed, as a political agenda seems to justify corrupted science to achieve the end. CRU people were involved from the start and triggered the first problems.

The leaked emails triggered a shock wave reflecting the degree to which climate science was politicized therefore it required the top political spin-doctors. University spokesperson Trevor Davies said it was a *reputation management* problem, which he claimed they don't handle well. Apparently, telling the truth was not considered.

George Monbiot of *The Guardian* actively sold the scientific rubbish produced by the IPCC whose major scientists were members of the CRU. This makes his reaction more telling. He was shocked by the emails and said, *Why was CRU's response to this issue such a total car crash?* The answer, George, is, because it was true and they were caught with their deceits and manipulations exposed. However, he made the bizarre argument that the transgressions were as nothing compared to the actions of the skeptics. Despite this his response is little short of floundering. They are those of someone who put far too much trust in what these people were saying because it suited his political bias. He wanted to believe. He wrote:

> Climate sceptics have lied, obscured and cheated for years. That's why we climate rationalists must uphold the highest standards of science.

He cites Hoggan and Littlemore's book *Climate Cover-Up* as an example of skeptic's behavior, I am one of the people attacked in the book and know the many errors, and lies about me and my career. The leaked emails exposed the liars and the manipulators, but Monbiot doesn't deal with it:

Chapter Ten

*But the deniers' campaign of lies, grotesque as it is, does
not justify secrecy and suppression on the part of climate
scientists. Far from it: it means that they must distinguish
themselves from their opponents in every way. No one has
been as badly let down by the revelations in these emails
as those of us who have championed the science. We
should be the first to demand that it is unimpeachable,
not the last.*

Investigations of Wrongdoing Exposed By Leaked Emails: More Cover-Up

Global warming science carried out by the IPCC was
premeditated. This was finally revealed in the 6,000 leaked
emails. I discussed previously how they diverged from the
standard practices of science that require scientists to disprove a
hypothesis. The emails provide the evidence of the methods used
to pursue what they referred to as "the cause." In their own words
they tell us how they created the 'scientific' evidence to support
the political agenda.

Proving the hypothesis required selecting or ignoring
evidence, or both. If these actions are exposed you have to admit
the error or try to cover up the truth. A response was required
after the emails were leaked that disclosed how climate science
was directed and distorted to achieve a premeditated result.

Pressure to restore confidence in the IPCC scientists was
intense and grew after the emails were leaked in November of
2009 and again in November 2011. Both achieved their objective
of ending Kyoto and any substitute. People like Monbiot were
rattled, but shrugged off the problems. Others, especially
politicians, began to ask questions.

In order to salvage the entire IPCC process it was essential to
begin by restoring credibility of the scientists exposed in the

leaked emails. The decision to pursue a cover-up was taken by major agencies directly involved, either because of political commitment or funding. They initiated five inquiries, but they were all orchestrated to mislead and cover up what was done.

Hans van Storch, Professor of Climatology wrote:

> We have to take a self-critical view of what happened. Nothing ought to be swept under the carpet. Some of the Inquiries—like—in the UK—did exactly the latter. They blew an opportunity to restore trust.[158]

Lord Turnbull summarized the serious allegations as follows in a Foreword to a study by Andrew Montford:

- Scientists at the CRU had failed to give a full and fair view to policymakers and the IPCC of all the evidence available to them
- They deliberately obstructed access to data and methods to those taking different viewpoints for themselves
- They failed to comply with Freedom of Information requirements
- They sought to influence the review panels of journals in order to prevent rival scientific evidence from being published.

---

[158] Von Storch, H. Wir mussen die Herausforderung durch die Skeptiker annehmen. Interview with Daniel Lingenhohl, Handelsblatt, August 2nd, 2010.

Chapter Ten

Three investigations in the United Kingdom included:

1. The UK House of Commons Science and Technology Committee, required because of the involvement of the United Kingdom Meteorological Office (UKMO) and its links with the CRU. A former Director of the UKMO, John Houghton was appointed first head of the IPCC.

   This inquiry seems pointless other than for politicians to be able to say they had considered the issue of the leaked emails. They initially canceled their inquiry when the University of East Anglia said they were investigating. Apparently because of pressure or because some saw political opportunities they sought input from the public and only received 58 submissions. They effectively deferred to the UEA. Sir Muir Russell, Chair of one of the UEA investigations appeared before them to explain what he was doing.

   Four other panels were questioned, but included nobody qualified or knowledgeable about climatology. This knowledge is essential to understand what the emails were saying. The Inquiry did not finish its work because of an election. They assumed the Oxburgh and Russell inquiries would resolve the matter. Attempts to strongly admonish the CRU for their actions were all defeated.

2. The Oxburgh panel was appointed and directed by the University of East Anglia (UEA). This the most compromised of the inquiries, so much so that its findings are meaningless except as evidence of a cover-up.

   There were no written terms of reference. UEA said Oxburgh was going to investigate the science. He didn't. Oxburgh and his committee of six were recommended by the Royal Society, which was directly involved in promoting the IPCC and CRU.

185

Oxburgh was compromised because he is a Fellow of the Royal Society, but more important, he is CEO of Carbon Capture and Storage Association and Chairman of Falck Renewable Resources that benefit from the claim that human CO_2 is causing warming. He also promoted the warming claim as UK Vice-Chair of GLOBE International, a consortium of Industry, NGOs' and Government that lobbies for global warming policy.

The Oxburgh Inquiry was directed to examine the CRU science, but failed. The other failures:

- There were no public hearings.
- There was no call for evidence.
- Only 11 academic papers were examined, a list vetted by Phil Jones, Director of the CRU the agency central that was being investigated.
- Only unrecorded closed interviews with CRU staff were held.
- There were no meetings with CRU critics.
- UEA had effective control of the Inquiry throughout.
- The UK House of Commons Select Committee grilled Oxburgh on the shallowness of his study and report and its failure to review the science as promised.

The Independent Climate Change Emails Review (ICCER), more commonly called the Muir Russell Inquiry, was also created by the University of East Anglia.

It was compromised from the start through conflict of interest of members. One person appointed was Philip Campbell, editor of *Nature*. He resigned when his bias was revealed. Another appointee, Geoffrey Boulton, had two major problems. He had signed a petition from the UK Met Office declaring full support for the CRU and IPCC science. He had been employed at UEA when the Inquiry said members had no links to the university. He said he was not a climate expert when a CV sent to a Chinese University stated the opposite.

There was a call for public submissions on February 11th, 2010 with a virtually impossible deadline of March 1st, 2010 (17 days). They did not hold public hearings and only interviewed CRU and UEA staff. Those items alone are sufficient to indicate the bias of the inquiry to a preconceived result. In a commentary on the Muir Russell Report, Fred Pearce of the *UK Guardian*, a paper long known for its strong support of the IPCC wrote:[159]

> *Secrecy was the order of the day at CRU. "We find that there has been a consistent pattern of failing to display the proper degree of openness," says the report. That criticism applied not just to Jones and his team at CRU. It applied equally to the university itself, which may have been embarrassed to find itself in the dock as much as the scientists on whom it asked Russell to sit in judgment.*

Montford's report showed all three Inquiries and their reports have serious flaws. Lord Turnbull summarized Montford's findings as follows:

- These inquiries were hurried
- In terms of reference were unclear
- Insufficient care was taken with the choice of panel members to ensure balance in independence

---

[159] http://www.theguardian.com/environment/cif-green/2010/jul/07/climategate-scientists

- Insufficient care was taken to ensure the process was independent of those being investigated, eg the Royal Society allowed CRU to suggest the papers it should read
- Sir Muir Russell failed to attend the session with the CRU's Director Professor Jones and only four of fourteen members of the Science and Technology Select Committee attended the crucial final meeting to sign off their report.
- Record keeping was poor.

Lord Turnbull concludes:

> But above all, Andrew Montford's report brings out the disparity between the treatment of the incumbents and the critics. The former appeared to be treated with kid gloves and their explanations readily accepted without serious challenge. The letter has been disbursed denied adequate opportunity to put their case.[160]

One investigation in North America at Penn State:

1. Penn State University appointed an inquiry because of involvement of Michael Mann, director of the Earth System Science Center at that university. Penn State University has procedures for Inquiry and then Investigation of academic wrong-doing. In the leaked emails case they carried out an Inquiry. The rules require five (5) tenured professors with competency but no conflict. The committee appointed comprised one tenured professor, one untenured with an MS in

---

[160] http://www.thegwpf.org/images/stories/gwpf-reports/Climategate-Inquiries.pdf

Psychology and an administrator. The untenured professor left during the Inquiry and was replaced by another administrator.

There was no call for evidence or public hearings. They only interviewed three people. Michael Mann was asked about questions of which he had prior notice. Gerald North of Texas A & M and Donald Kennedy of Stanford were interviewed. Neither was involved with the emails, but were publicly sympathetic to the IPCC work.

Comments by Clive Crook about the Penn State Inquiry provide an excellent summary:

> The Penn State inquiry exonerating Michael Mann—the paleoclimatologist who came up with the hockey stick—would be difficult to parody. Three of four allegations are dismissed out of hand at the outset: the inquiry announces that, for lack of credible evidence, it will not even investigate them. (At this, MIT's Richard Lindzen tells the committee, "It's thoroughly amazing. I mean these issues are explicitly stated in the emails. I'm wondering what's going on?" The report continues: "The Investigatory Committee did not respond to Dr. Lindzen's statement. Instead his attention was directed to the fourth allegation.") Moving on, the report then says, in effect, that Mann is a distinguished scholar, a successful raiser of research funding, a man admired by his peers—so any allegation of academic impropriety must be false.

Steve Milloy, founder of the web page *JunkScience*, explains why Crook was so dissatisfied did a similar analysis of the Penn State Inquiry as Andrew Montford:

> 1. The review apparently extended little further than the Climategate e-mails themselves, an interview with Mann, materials submitted by Mann and whatever e-mails and

comments floated in over the transom. Not thorough at all.

2. Comically, the report explains at length how the use of the word "trick" can mean a "clever device." The report ignores that it was a "trick...to hide the decline." There is no mention of "hide the decline" in the report.

3. The report concludes there is no evidence to indicate that Mann intended to delete e-mails. But this is contradicted by the plain language and circumstances surrounding Mann's e-mail exchange with Phil Jones— See page 9 of Climategate & Penn State: The Case for an Independent Investigation.[161]

4. The report dismisses the accusation that Mann conspired to silence skeptics by stating, "one finds enormous confusion has been caused by interpretations of the e-mails and their content." Maybe there wouldn't be so much "confusion" if PSU actually did a thorough investigation rather than just relying on the word of Michael Mann.

5. Although PSU is continuing the investigation, its reason is not to investigate Mann so much as it is to exonerate climate alarmism. On page 9 of the report, it says that "questions in the public's mind about Dr. Mann's conduct...may be undermining confidence in his findings as a scientist...and public trust in science in general and climate science specifically."[162]

---

[161]

http://www.commonwealthfoundation.org/docLib/20100111_PB220 2Climategate.pdf)

[162] http://www.thestreet.com/story/10673791/1/milloy-comments-on-penn-state-scandal-and-investigation-of-michael-mann.html

Chapter Ten

Result of One International investigation

The Inter Academy Council (IAC). This is a UN created group designed to act as a public relations panel for national academies of science. It was commissioned by the IPCC to investigate their procedures. It is partly limited because of previous close and conflicting connections with IPCC Chair, Rajendra Pachauri.[163]

A Brief Overview of the Inquiries

The brief analysis of each inquiry explains why Clive Crook, Senior editor of The Atlantic wrote:

> *I had hoped, not very confidently, that the various Climategate inquiries would be severe. This would have been a first step towards restoring confidence in the scientific consensus. But no, the reports make things worse. At best they are mealy-mouthed apologies; at worst they are patently incompetent and even willfully wrong. The climate-science establishment, of which these inquiries have chosen to make themselves a part, seems entirely incapable of understanding, let alone repairing, the harm it has done to its own cause.*

George Monbiot, another *Guardian* reporter and a fierce advocate of the IPCC and the CRU who is part of the leaked emails wrote on his personal blog,[164]

> *It's no use pretending that this isn't a major blow. The emails extracted by a hacker from the climatic research unit at the University of East Anglia could scarcely be*

---

[163] http://www.climatescience.gov/Library/bios/pachauri.htm

[164] http://www.monbiot.com/2009/11/23/the-knights-carbonic/

more damaging. I am now convinced that they are genuine, and I'm dismayed and deeply shaken by them. Yes, the messages were obtained illegally. Yes, all of us say things in emails that would be excruciating if made public. Yes, some of the comments have been taken out of context. But there are some messages that require no spin to make them look bad. There appears to be evidence here of attempts to prevent scientific data from being released, and even to destroy material that was subject to a Freedom of Information request.

Worse still, some of the emails suggest efforts to prevent the publication of work by climate sceptics, or to keep it out of a report by the Intergovernmental Panel on Climate Change. I believe that the head of the unit, Phil Jones, should now resign. Some of the data discussed in the emails should be re-analysed.

Overall Summary of the Investigations

There was a distinct pattern to the process used in each inquiry, which was clearly dictated by the cover-up objective.

- The people appointed to the inquiries were either compromised through conflict or had little knowledge of climatology or the IPCC process.
- They did not have clearly defined objectives and failed to achieve any they publicized.
- Interviews were limited to the accused.
- Experts who knew what went on and how it was done, that is understood what the emails were saying, were not interviewed.
- Validity of the science and the results obtained as published in the IPCC Reports were not examined, yet the deceptions were to cover these problems.

- All investigations were seriously inadequate in major portions so as to essentially negate their findings. It appears these inadequacies were deliberate to avoid exposure of the truth. They all examined only one limited side of the issues, so it was similar to hearing only half of a conversation and what you hear is preselected.

Those involved in the cover-up achieved their goal because the media stopped asking questions. It also allowed those identified in the emails to claim they were absolved of any wrong-doing.[165]

As Emeritus Professor Garth Paltridge said:

> *Basically, the problem is that the research community has gone so far along the path of frightening the life out of the man in the street that to recant publicly even part of the story would massively damage the reputation and political clout of science in general. And so, like corpuscles in the blood, researchers all over the world now rush in overwhelming numbers to repel infection by any idea that threatens the carefully cultivated belief in climatic disaster.*[166]

---

[165] http://blogs.discovermagazine.com/intersection/tag/michael-mann/#.UkX-KxbVl38

[166] http://www.lavoisier.com.au/articles/climate-policy/science-and-policy/paltridge2008-13.php

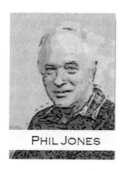

PHIL JONES

CHAPTER ELEVEN

What Else Did We Learn From the Leaked Emails?

GEORGE MONBIOT OF *The Guardian* (UK) was among the first mainstream media to express concern about the content of some leaked emails and the story they told:

> *I am now convinced that they are genuine, and I'm dismayed and deeply shaken by them.*[167]

He was reacting to corruption on an unprecedented scale exposed in the leaked emails. It appears he was only concerned about being fooled because he later appeared in denial of the extent of the deception. He appears like most unable to believe what was done. People generally find it hard to believe the extent to which a few scientists controlled climate science. It is likely encompassed by the adage that no one person or even a small group can change the world. It began with limiting the research and researchers involved with the IPCC. Then it reduced to the people at the

---

[167] http://www.monbiot.com/2009/11/23/the-knights-carbonic/

CRU and a few trusted associates. We have a specific measure of most of them.

Individuals and groups joined and for various reasons most were unaware of the overall objective laid out by Maurice Strong, the Club of Rome and then UNEP. Many scientists became involved because of research funding. The funding and a political agenda attracted a few. A majority of people and groups participated because of the new paradigm of environmentalism. All the attractions made people vulnerable, as Strong and others knew. It was a classic exploitation of human nature, especially the weaknesses.

The U.S. Congress Chairman of the Committee on Energy and Commerce and the Chairman of the Subcommittee on Oversight and Investigations appointed Professor Wegman to head an independent investigation of the "hockey stick" dispute:

> *In general, we found MBH98 and MBH99 [the original hockey stick papers by Mann, Bradley and Hughes] to be somewhat obscure and incomplete and the criticisms of MM03/05a/05b (by McIntyre and McKitrick) to be valid and compelling.*

However, in a most remarkable and detailed analysis the report found:

> *In our further exploration of the social network of authorships in temperature reconstruction, we found that at least 43 authors have direct ties to Dr. Mann by virtue of coauthored papers with him. Our findings from this analysis suggest that authors in the area of paleoclimate studies are closely connected and thus 'independent studies' may not be as independent as they might appear on the surface.*

Wegman went further:

> *It is important to note the isolation of the paleoclimate community; even though they rely heavily on statistical methods they do not seem to be interacting with the statistical community. Additionally, we judge that the sharing of research materials, data and results was haphazardly and grudgingly done. In this case we judge that there was too much reliance on peer review, which was not necessarily independent.*

Michael Mann couldn't understand the fuss:

> *"What they've done is search through stolen personal emails—confidential between colleagues who often speak in a language they understand and is often foreign to the outside world," Penn State's Michael Mann told Reuters Wednesday. Mr. Mann added that this has made "something innocent into something nefarious."*[168]

The key comment here is *foreign to the outside world*. That is the obscurity they relied on in their larger public dealings. They knew few would understand and they could marginalize those who did and dared to point it out:

> *Phil Jones, director of the CRU appeared to have a different view. "My colleagues and I accept that some of the published emails do not read well. I regret any upset or confusion caused as a result. Some were clearly written in the heat of the moment, others use colloquialisms*

---

[168]

http://online.wsj.com/article/SB1000142405274870349940457455963038 2048494.html *frequently used between close colleagues…"*

It is a partial excuse, but relies on most not understanding. This chapter will detail what they said and how it was rarely in the heat of the moment. The only heat occurred when somebody did or said something they didn't like.[169]

Some still deny the extent and impact of what the CRU and their supporters did. Although fully orchestrated, most participated unaware of the scientific deception, lured by funding, employment and sometimes prestige. A few saw it as part of the big green lie that has pushed environmentalism.[170] Others saw it is as a big lie on its own.[171] They knew exactly what they were doing because if they were dealing with the truth they wouldn't have needed the Palace Guard they created.

True, the scale and extent appears unbelievable. However, I know it's believable because I watched it develop and grow. Particularly since 1985 when the conference in Villach Austria conjoined the CRU with the fledgling Intergovernmental Panel on Climate Change (IPCC).

Tom Wigley and Phil Jones, both Directors of the CRU, attended but were already developing the phony climate science Maurice Strong needed to pursue his goal of destroying western economies. For example, in a 1983 article Wigley was convincing climate science of a falsely low pre-industrial level of CO_2. Early attempts to challenge what they were doing followed normal academic processes with no effect. For example, I wrote a book review based on the bad science.[172]

---

[169] Ibid.

[170] http://blogs.the-american-interest.com/wrm/2010/07/12/the-big-green-lie-exposed/

[171] http://euro-med.dk/?p=15111

[172] This review became a Review Editorial in *Climatic Change* (Volume 35, Number 4 / April, 1997.)

Wigley was central to what went on. He took over the CRU from Hubert Lamb who reports in his autobiography, *Through All the Changing Scenes of Life* that he set it up because:

> ...it was clear that the first and greatest need was to establish the facts of the past record of the natural climate in times before any side effects of human activities could well be important.

There was a conflict almost immediately. As Lamb wrote:

> The research project which I put forward to the Rockefeller Foundation was awarded a handsome grant, but it sadly came to grief over an understandable difference of scientific judgment between me and the scientist, Dr. Tom Wigley, whom we appointed to take charge of the research...
>
> The scheme had been to extract the information given in the wealth of descriptive reports of the nature of individual past seasons...
>
> My plan was that these reports should be entered on maps of the reported weather character that prevailed in the individual seasons...

This conflict is central to the problem with climate research today. We have less data now than when Lamb identified the problem and mostly due to the shift in ideology and emphasis. Lamb identified the problem with Wigley and ultimately the CRU and the IPCC:

> My immediate successor, Professor Tom Wigley, was chiefly interested in the prospects of world climates being changed as result of human activities, primarily through the burning up of wood, coal, oil and gas reserves...After

only a few years almost all the work on historical
reconstruction of past climate and weather situations,
which first made the Unit well known, was abandoned.

Wigley and his graduate students were instrumental in the application of computer models, but as Lamb knew they were only as good as the data on which they were built. They were and continue to be a disaster, while Lamb's work and the studies it engendered, prove prescient.

The big change came when computer modelers took over climate science. They provided the ability to deal with large volumes of data for basic statistical analysis, but they were primarily focused on weather and later climate forecasting There were earlier simple numerical models but the real impact came in the 1970s with the work of Syukuro Manabe. Bert Bolin, first Co-chair of the IPCC, provided an early warning about the limitations when he wrote:

There is very little hope for the possibility of deducing a
theory for the general circulation of the atmosphere from
the complete hydrodynamic and thermodynamic
equations. [173]

This statement is still essentially true and part of the debate about climate-based strategy. In 1977, Abelson wrote about more apparent limitations:

Meteorologists still hold out global modeling as the best
hope for achieving climate prediction. However, optimism
has been replaced by a sober realization that the problem

---

[173] Bolin, Bert (1952). "Studies of the General Circulation of the Atmosphere." *Advances in Geophysics* 1: 87-118.

is enormously complex.[174]

I knew modeling global climate was impossible; apart from the inadequate surface and upper atmosphere database, computer capacity was and is still inadequate. At conference after conference, I watched modelers bully everybody. Models are the most corrupt part of the CRU and IPCC fiasco, an exposure yet to emerge. They produced the ridiculous 'scenarios' of disaster used to promote control through fear.

We've learned of data manipulation, publication and peer review control, and personal attacks on those who asked questions. Still to fully emerge is how they manipulated the computer models to reach a result that was not a simulation of nature, but proof that human CO_2 was causing global warming and climate change. As the IPCC and its model projections grew in power to dominate global energy policy, it drew increasing attention. This grew threatening and triggered the need for a Palace Guard to defend the CRU and thereby the IPCC. The emails gave insights into the structure, objectives and methods.

The Palace Guard

A group of scientists established themselves as the palace guard for the bosses at the CRU. Michael Mann and Gavin Schmidt led and quickly earned reputations for nasty and vindictive responses. On December 10th, 2004 Schmidt gave the CRU gang a Christmas present:

> *Colleagues, No doubt some of you share our frustration*
> *with the current state of media reporting on the climate*
> *change issue. Far too often we see agenda-driven*
> *"commentary" on the Internet and in the opinion columns*

---

[174] Abelson, P.H. (1977). "Energy and Climate." *Science* 197: 941.

of newspapers crowding out careful analysis. Many of us work hard on educating the public and journalists through lectures, interviews and letters to the editor, but this is often a thankless task. In order to be a little bit more pro-active, a group of us (see below) have recently got together to build a new 'climate blog' website: RealClimate.org which will be launched over the next few days.

The group was Mike Mann, Eric Steig, William Connolley, Stefan Rahmstorf, Ray Bradley, Amy Clement, Rasmus Benestad and Caspar Ammann—they're familiar names to people who crossed them. Now the world knows. Evasiveness pervading the behavior recorded in the CRU emails was present at RealClimate (RC) and beyond.

Further, Schmidt elaborated:

The idea is that we working climate scientists should have a place where we can mount a rapid response to supposedly 'bombshell' papers that are doing the rounds and give more context to climate related stories or events.

The phrase *working climate scientists* frequently appears and typifies their arrogance. Unless you are one, you have no credibility or right to an opinion. It's similar to their peer review charge and establishes them as the elite.

On October 23rd, 2003, Ray Bradley illustrated how the group could defend itself through asserting moral and intellectual superiority.

Because of the complexity of the arguments involved, to an uninformed observer it all might be viewed as just scientific nit-picking by "for" and "against" global warming proponents. However, if an "independent group" such as you guys at CRU could make a statement as to whether the M&M (McIntyre and McKitrick) effort is truly an "audit", and if they say it right, I think would go a long way to defusing the issue. If you are willing, a quick and forceful statement from The Distinguished CRU Boys would help quash further arguments, although here, at least, it is already quite out of control.

The Modus Operandi Involved the Mainstream Media

Activities of these *working climate scientists* were not to answer questions about their work but to divert, distract, ignore and marginalize with lies about people and ideas. As Thomas Chase said:

> *The single biggest scandal revealed in the emails from the Climatic Research Unit is the lengths they went to refuse outside requests to make data and methodology available over the course of years including discussions about resisting Freedom of Information Act requests.*[175]

Here is a February 9th, 2006 email from Michael Mann that gives a flavor of this behavior.

> *I see that Science (the journal) has already gone online w/ the new issue, so we put up the RC post. By now, you've probably read that nasty McIntyre thing. Apparently, he violated the embargo on his website (I*

---

[175]

http://www.dailycamera.com/archivesearch/ci_14151309?IADID=Se arch-www.dailycamera.com-www.dailycamera.com

don't go there personally, but so I'm informed). Anyway, I wanted you guys to know that you're free to use RC in any way you think would be helpful. Gavin and I are going to be careful about what comments we screen through, and we'll be very careful to answer any questions that come up to any extent we can. On the other hand, you might want to visit the thread and post replies yourself. We can hold comments up in the queue and contact you about whether or not you think they should be screened through or not, and if so, any comments you'd like us to include. You're also welcome to do a follow-up guest post, etc. think of RC as a resource that is at your disposal to combat any disinformation put forward by the McIntyres of the world. Just let us know. We'll use our best discretion to make sure the skeptics don't get to use the RC comments as a megaphone...

Mann spread his views about McIntyre through Andrew Revkin of the New York Times. As recently as September 29th, 2009 he wrote:

...those such as McIntyre who operate almost entirely outside of this system are not to be trusted.

Jones did it when he defended his refusal to answer FOI's to the administration at the University of East Anglia. The emails from Revkin are disturbing and reveal unhealthy involvement and lack of journalistic integrity. No wonder he blocked use of the Climategate material in the newspaper when it appeared. It was not journalistic integrity; it covered his involvement. He finally resigned from the newspaper, but it appears to be a move to avoid the heat because he continues to produce material through a blog site and occasional articles in the NYT.

Schmidt notes:

> *This is a strictly volunteer/spare time/personal capacity*
> *project and obviously nothing we say there reflects any*
> *kind of 'official' position.*

What hypocrisy, they were the official position. Many argue it wasn't very "volunteer/spare time" because Schmidt worked full time at NASA GISS. This is the game James Hansen and others play. He is Director of the NASA's Goddard Institute of Space Studies (GISS) when it suits and a private citizen when it suits. It's a duplicity that underlines the political nature of their activities.

Wikipedia—a Falsified Resource for Students and Media

Perhaps the most insidious activity included controlling climate information through Wikipedia. When I ask students how many use Wikipedia for their research all hands go up. I know most media rely on it. Most don't know how the material was entered or edited. They are learning partly due to the activities of people connected to the CRU group.

William Connolley knew and exploited the opportunity.[176] A participant in computer modeling, his activities are shocking.[177] He established himself as an editor at Wikipedia and with a cadre (I use the term deliberately) of supporters he controlled all entries relating to climate, climate change and the people involved. This included putting up false material about skeptics. They constantly monitored the entries and rapidly returned to the original false information any attempted corrections. With so many people, they could easily circumvent the limit on the number of edits per

---

[176] http://wattsupwiththat.com/2009/12/19/wikibullies-at-work-the-national-post-exposes-broad-trust-issues-over-wikipedia-climate-information/

[177] http://www.conservapedia.com/William_M._Connolley

person. Connolley as a designated editor had even more latitude. Here is how Lawrence Solomon described the activities:

> *All told, Connolley created or rewrote 5,428 unique Wikipedia articles. His control over Wikipedia was greater still, however, through the role he obtained at Wikipedia as a website administrator, which allowed him to act with virtual impunity. When Connolley didn't like the subject of a certain article, he removed it—more than 500 articles of various descriptions disappeared at his hand. When he disapproved of the arguments that others were making, he often had them barred—over 2,000 Wikipedia contributors who ran afoul of him found themselves blocked from making further contributions. Acolytes whose writing conformed to Connolley's global warming views, in contrast, were rewarded with Wikipedia's blessings. In these ways, Connolley turned Wikipedia into the missionary wing of the global warming movement.*

The Medieval Warm Period disappeared, as did criticism of the global warming orthodoxy. With the release of the Climategate Emails, the disappearing trick has been exposed. The glorious Medieval Warm Period will remain in the history books, perhaps with an asterisk to describe how a band of zealots once tried to make it disappear.

Wikipedia eventually removed Connolley as editor, but not permanently as he later was reinstated only to continue his activities.[178] His actions underline the level of control and misdirection the CRU people and the Palace Guard practiced. As the German magazine *Die Kalte Sonne* reported in January 2013:

---

[178] http://en.wikipedia.org/wiki/William_Connolley

Unbelievable but true: The Wikipedia umpire on Climate Change was a member of the UK Green Party and openly sympathized with the views of the controversial IPCC. So it was not a referee, but the 12th Man of the IPCC team.[179]

Control of the Peer Review Process

I became suspicious when people were classified and derided by the Palace Guard and others for the number of "peer reviewed" articles they published. It was an extension of the attempt to isolate themselves as an elite group. They were the only people qualified and it is similar to their use of phrases like, *working climatologist* or *active climatologist*. These appear to be part of the overall PR campaign, but more important, they make real, but frightening, sense when you know they knew they controlled publication of climate papers.

This elitism and bias was reflected in the media. Whenever I and other "skeptics" appear on radio or TV the first question usually asked is about qualifications. I rarely heard this asked of pro-AGW guests. It certainly was never asked of Al Gore. I was qualified and could explain the issues in ways people could understand and that made me a big target.

A major part of the control of research occurred when Maurice Strong established the IPCC through the WMO. This put the national weather agencies in control of research because funding in most countries went through them. These agencies essentially only funded research that supported the IPCC view.

The second part of the control was more direct. For example, Donna Laframboise, author of the devastating book

---

[179]

http://translate.google.com/translate?sl=auto&tl=en&js=n&prev=_t&hl=en&ie=UTF-8&eotf=1&u=http://www.kaltesonne.de/?p=7858

about the corruption of the IPCC called *The Delinquent Teenager,* lists all the people serving on the IPCC who were editors or on the editorial board of the *Journal of Climate.*[180] The opening comment reads:

> *We're supposed to trust the conclusions of the Intergovernmental Panel on Climate Change (IPCC) because much of the research on which it relies was published in peer-reviewed scientific journals. But what happens when the people who are in charge of these journals are the same ones who write IPCC reports?*

The idea of the predominance of one side of the climate debate was reinforced by a 2004 publication by Naomi Oreske. It claimed that of 928 publications obtained in an Internet search, all supported the IPCC's AGW hypothesis. There is no point elaborating because the fallacies and bias in the research were soon identified. Peer reviewed papers contradicting Oreske's claims were quickly produced.[181] The most damning of her work was the comment by Tom Wigley, former Director of the CRU and central player in the climate science manipulation. On November 12th, 2009 he wrote to Phil Jones about her work saying, *analyses like these by people who don't know the field are useless. A good example is Naomi Oreskes' work.*[182]

When the emails were leaked from the Climatic Research

---

[180] http://nofrakkingconsensus.com/2011/08/23/the-journal-of-climate-the-ipcc/

[181] http://www.skepticalscience.com/naomi-oreskes-consensus-on-global-warming.htm

[182] http://junkscience.com/2013/03/13/climategate-3-0-tom-wigley-says-naomi-oreskes-work-is-useless/comment-page-1/

Unit (CRU) a public relations person was engaged.[183] Involvement of PR people is evident in almost everything. The web site DeSmogBlog is the brainchild of James Hoggan, Board Chairman of the David Suzuki Foundation and President of a PR firm.[184] In a December 2011 email to Michael Mann, DeSmogBlog writer Richard Littlemore says:

> (as I am sure you have noticed: we're all about PR here, not much about science).[185]

Evidence it's a PR battle is Mike Mann's 2004 email to CRU Director Phil Jones.[186] Confronted by challenging questions, they apparently developed a defensive mentality:

> I've personally stopped responding to these, they're going to get a few of these op-ed pieces out here and there, but the important thing is to make sure they're loosing [sic] the PR battle. That's what the site is about. By the way, Gavin did come up w/ the name!

The "site" is the web site RealClimate, set up by the Palace Guard and named by Gavin Schmidt. Science doesn't need PR, so why do climate scientists need it?

The PR battle involved proving their superiority in credibility, rigour and quality of published work. A March 2003 email from Mann to Jones proves that exploitation of peer review

---

[183] http://climateaudit.org/2011/07/14/covert-operations-by-east-anglias-cru/

[184] http://www.hoggan.com/

[185] http://tomnelson.blogspot.ca/2011/12/desmogblog-climate-hoax-promoter.html

[186]

http://www.ecowho.com/foia.php?file=1485.txt&search=PR+battle

was a deliberate strategy:

> *This was the danger of always criticising the skeptics for
> not publishing in the "peer-reviewed literature".*

As one university library tells students:

> Peer review ensures that an article—and therefore the
> journal and the scholarship of the discipline as a
> whole—maintains a high standard of quality, accuracy,
> and academic integrity.[187]

Are CRU the only arbiters of quality and accuracy? Apparently, as
Mann said in a March 2003 email:

> *The Soon & Baliunas paper couldn't have cleared a
> 'legitimate' peer review process anywhere. That leaves
> only one possibility—that the peer-review process at
> Climate Research has been hijacked by a few skeptics on
> the editorial board. And it isn't just De Frietas,
> unfortunately I think this group also includes a member
> of my own department…The skeptics appear to have
> staged a 'coup' at "Climate Research" (it was a mediocre
> journal to begin with, but now it's a mediocre journal
> with a definite 'purpose'). Obviously, they found a
> solution to that—take over a journal![188]*

They weren't finished:

> *So what do we do about this? I think we have to stop*

---

[187] http://library.uvic.ca/instruction/research/peerreview.html
[188]

http://www.ecowho.com/foia.php?file=1047388489.txt&search=pee
r+review

> considering "Climate Research" as a legitimate peer-reviewed journal. Perhaps we should encourage our colleagues in the climate research community to no longer submit to, or cite papers in, this journal. We would also need to consider what we tell or request of our more reasonable colleagues who currently sit on the editorial board...

The irony in their view of a group controlling journals is that they don't seem to realize that it is precisely what they are doing.

The control of peer reviewed publications took two forms. The first involved guaranteeing publication of their own work by peer reviewing each other's papers and only working with journals sympathetic to their view. This completely negates the major objective of peer review, which is to ensure rigorous application of the scientific method. The second, involved prevention of publications that contradicted, challenged, or openly disclosed inadequacies of data and method.

The centerpiece of their proof that humans were causing global warming was the hockey stick graph. Because the IPCC ostensibly only used peer reviewed papers, the research was available before it became central to the 2001 IPCC Science report and Summary for Policymakers. It appeared in *Nature* 392, 779-787 (April 23rd,1998) authored by Michael E. Mann, Raymond S. Bradley and Malcolm K. Hughes, *Global-scale temperature patterns and climate forcing over the past six centuries* usually abbreviated to MBH98.[189] Steve McIntyre and Ross McKitrick recognized the problems with the hockey stick, which they identified as a misuse of a statistical technique and how the results were due to methodological artifact. The latter meant that

---

[189]

http://climateaudit101.wikispot.org/Mann,_Bradley_&_Hughes_(199 8)

the hockey stick shape resulted, even if random data was put into the grinder. Their first analysis appeared in a Dutch journal *nwtonline* in 2003 and drew some response.[190] Dr. Rob van Dorland, an IPCC Lead author made the noteworthy comment that their analysis *seriously damaged the image of the IPCC*. More important, he said:

> ...*it is strange that the climate reconstruction of Mann has passed both peer review rounds of the IPCC without anyone ever really having checked it. I think this issue will be on the agenda of the next IPCC meeting in Peking this May.*

I don't think it made the agenda because they try to defend the hockey stick.

A second article appeared in *Energy and Environment* in 2003[191], but it was an article published in *Geophysical Research Letters (GRL)* that caused the CRU gang and the Palace Guard the greatest consternation. It became a topic of discussion between Michael Mann, Malcolm Hughes and Tom Wigley on January 21st, 2005. The themes were a concern about their inability to control publications in GRL and how to respond and limit the damage. In another email Michael Mann comments:

> ...*apparently, the contrarians now have and "in" with GRL. This guy Saiers (editor) has a prior connection w/ the University of Virginia Department of environmental sciences that causes me some unease.*

He is apparently referring to Patrick Michaels, a long-term skeptic resident there at that time. Mann continues:

---

[190] http://www.nwtonline.nl/index.html
[191] http://www.multi-science.co.uk/mcintyre-mckitrick.pdf

I'm not sure that GRL can be seen as an honest broker in these debates anymore and it is probably best to do an end run around GRL now where possible. They have published far too many deeply flawed contrarian papers in the past year or so. There is no possible excuse for them publishing all 3 Douglass papers and the Soon et al paper. These were all pure crap.

...Basically this is just a heads up to people that something might be up here. What a shame that would be. It's one thing to lose "Climate Research". We can't afford to lose GRL. I think it would be useful if people begin to record their experiences w/ both Saiers and potentially Mackwell (I don't know him-he would seem to be complicit w what is going on here). If there is a clear body of evidence that something is amiss, it could be taken through the proper channels. I doubt that the entire page EU hierarchy has yet been compromised!

In a direct threat Phil Jones wrote:

I will be emailing the journal to tell them I'm having nothing more to do with it until they rid themselves of this troublesome editor.[192]

They apparently forced the resignation of the editor James Saiers.[193] A later email reported:

---

[192]

http://www.ecowho.com/foia.php?file=1047388489.txt&search=peer+review

[193]

http://www.americanthinker.com/2011/09/the_warmists_strike_back.html

The GRL leak may have been plugged up now w/new editorial leadership there.[194]

As Wesley J. Smith summarized at *National Review Online*:

Most disturbing is the major effort made by a small group of very powerful scientists to prevent skeptics from being published in peer reviewed journals and attacks on editors that published opposing views—sometimes costing the editors their jobs. Talk about making sure you have no competition![195]

The climate cabal went further by determining who would be a reviewer and what they would say:

Not content to block out all dissent from scientific journals, the CRU scientists also conspired to secure friendly reviewers who could be counted on to rubber-stamp their own work. Phil Jones suggests such a list to Kevin Trenberth, with the assurance that "All of them know the sorts of things to say...without any prompting."[196]

Other comments indicate the malfeasance. On June 4th, 2003

---

[194]

http://www.realclearpolitics.com/articles/2009/11/24/the_fix_is_in_99280.html

[195] http://www.nationalreview.com/human-exceptionalism/326203/global-warming-hysteria-damning-der-spiegel-expose-corruption-science-cl

[196]

http://www.realclearpolitics.com/articles/2009/11/24/the_fix_is_in_99280.html

Edward Cook asks Keith Briffa for help in rejecting an article he's reviewing. He writes:

> *Now something to ask from you. Actually somewhat important too. I got a paper to review (submitted to the Journal of Agricultural, Biological, and Environmental Sciences), written by a Korean guy and someone from Berkeley, that claims that the method of reconstruction that we use in dendroclimatology (reverse regression) is wrong, biased, lousy, horrible, etc. They use your Tornetrask recon as the main whipping boy. If published as is, this paper could really do some damage.*

Phil Jones emails Steve [Schneider], editor of Climatic Change, telling him he shouldn't accede to McIntyre's request for Mann's computer code:

> *Jones to Santer Mar 19th, 2009: I'm having a dispute with the new editor of Weather. I've complained about him to the RMS Chief Exec. If I don't get him to back down, I won't be sending any more papers to any RMS journals and I'll be resigning from the RMS.*
>
> *Note: Weather is a journal published by the Royal Meteorological Society (RMS).*

Santer replies:

> *If the RMS is going to require authors to make ALL data available—raw data PLUS results from all intermediate calculations—I will not submit any further papers to RMS journals.*

This is the crux of so much of the debate the failure of CRU and related people to provide data when they publish studies. The ability to test, known as reproducible results, is essential to science.

IPCC Report rules say they only consider peer reviewed articles. IPCC Chairman Rajendra Pachauri bragged:

> *The process is so robust—almost to a fault—that I'm not sure there is too much scope for error. Where there are gaps we are very candid in admitting we don't know enough about this subject...Given that it is all on the basis of peer-reviewed literature. I'm not sure there is any better process that anyone could have followed.*[197]

This was a false claim. A study of the 2007 Report found:

> *Forty citizen auditors from 12 countries examined 18,500 sources cited in the report—finding 5,600 to be not peer-reviewed.*[198]

Peer review became an incestuous system as they published together and apparently peer reviewed each other's work, as the Wegman Report identified.[199] It was critical in the public relations battle, but peer review when incestuous, works against innovation and perpetuates prevailing wisdoms. Editors can practice peer review censorship by selecting high priests of the prevailing wisdom to review and reject articles they consider heresy. Secrecy made the peer review process, that the CRU emails disclosed, possible. Editors weren't required to disclose reviewer's names. Where and when did that policy begin? There's no reason for secrecy and it contradicts a fundamental tenant of law; the right to

---

197

http://www.nzherald.co.nz/business/news/article.cfm?c_id=3&objectid=10514468&pnum=0

[198] http://www.noconsensus.org/ipcc-audit/press-release.php

[199] http://www.uoguelph.ca/~rmckitri/research/WegmanReport.pdf

face your accuser.

Summary of Peer Review Activities

- Control of the peer review process by reviewing each other's work
- Control by having a list of favorable reviewers provided to editors
- Control of peer review by threatening, isolating, or even eliminating journal editors
- Control of peer review by serving as editors of a significant climate journal
- Control of peer review by blocking publication of articles that question or reject their "science"
- Control of peer review, by a PR campaign to suggest only they, as "working climatologists", had credibility
- Ridiculed journals they considered inferior that published articles they didn't like

The final comment goes to Phil Jones, Director of the CRU. His comment confirms overall control of publication, but more important, their control over what appeared in the IPCC Reports:

> *I can't see either of these papers being in the next IPCC report. Kevin and I will keep them out somehow—even if we have to redefine what the peer-review literature is!*

Leaked Computer Information

All of the public attention was on the leaked emails and mostly on the first 1,000. They were enough to expose the corruption, so attention quickly waned about the detail. The second set of 5,000, released a year later drew little attention, yet contained much more valuable confirming and damning information. However, before we examine them there was a much bigger omission and that is the release of computer code in a file called *Harry Read Me*.[200] Elsewhere I discussed the use of computer models to predetermine the results that supported their hypothesis that human CO_2 is driving warming and climate change. It is a classic circular argument. Say an increase in CO_2 causes a temperature increase. Program the computer accordingly, then, when the results show temperature increasing, use it as proof of the hypothesis.

We know the computer results have been consistently wrong and it is likely due to what they put in and leave out, but we don't know. The computer codes used for the IPCC models are generally not available. The models are mathematical representations of each component of the climate system that are then expressed in computer terminology (code). The problem is each model is using different numbers for the same variable. For example, here is one author writing about the very important determination of climate sensitivity; the measure of how much temperature increase occurs when increasing CO_2:

> *Bizarrely, rather than use common forcings, each modeling group was directed to use forcings that it "deemed appropriate" so the various modeling groups are not using the same assumptions for the model inputs. Each group is modeling a different virtual planet.*[201]

It is not surprising that in a discussion about models,[202] Tamsin

---

[200] http://www.anenglishmanscastle.com/HARRY_READ_ME.txt

[201] http://www.climateviews.com/Climate_Views/Download_Articles_files/poster1iSmTit.pdf

[202] http://allmodelsarewrong.com/

Edwards concludes:

> *Models are always wrong, but what is more important is to know how wrong they are: to have a good estimate of the uncertainty about the prediction. Mark and Patrick explain that our uncertainties are so large because climate prediction is a chain of very many links.*

Knowledge of the computer codes is important, which is why they were not readily available. The leaked emails tell us about the concern to protect them. It comes up in a discussion about the Freedom of Information Act (FOIA). The University of East Anglia distributed a brochure in January 2005 with instructions about the Freedom of Information requests that they and the Climatic Research Unit were receiving. Tom Wigley had asked for answers from Phil Jones who replied:

> *On the FOI Act there is a little leaflet we have all been sent. It doesn't really clarify what we might have to do re programs or data. Like all things in Britain we will only find out when the 1ˢᵗ person or organization asks. I wouldn't tell anybody about the FOI Act in Britain. I don't think UEA really knows what's involved. As you're no longer an employee I would use this argument if anything comes along. I think it is supposed to mainly apply to issues of personal information references for jobs...*[203]

One author understood that Jones...

---

[203]

http://www.assassinationscience.com/climategate/1/FOIA/mail/110
6338806.txt

...certainly seems to be looking for ways to hide from complying with the law rather than complying with it. Wigley is no better in his response either in his understanding of the law or his willingness to try to find ways to not comply with it.

He writes back to Jones expressing his concern about the release of the computer codes:

Thanks for the quick reply. The leaflet appeared so general, but it was prepared by UEA so they may have simplified things. From their wording, computer code would be covered by the FOIA. My concern was if Sarah is/was still employed by UEA. I guess she could claim that she had only written one 10th of the code and release every 10th line.

Wigley knew what was in the computer codes, the Harry Read Me files. Jones replied with words to comfort Wigley, but very discomforting for anyone concerned about ethics in science:

As for FOIA Sarah isn't technically employed by the UEFA and she likely will be paid by Manchester Metropolitan University. I wouldn't worry about the code. If FOIA does ever get used by anyone, there is also IPR (Intellectual Property Rights) to consider as well. Data is covered by all the agreements we signed with people, so I will be hiding behind them. I'll be passing any requests on to the person at UEA who has been given a post to deal with them.

What they forget is the work they are doing is paid for by the taxpayer and used by politicians to have an impact on those taxpayers' lives.

In his report on the hockey stick debacle for the U.S. House of Representatives Energy and Commerce Committee Professor Wegman wrote:

> *Sharing of research materials, data, and results is haphazard and often grudgingly done. We were especially struck by Dr. Mann's insistence that the code he developed was his intellectual property and that he could legally hold it personally without disclosing it to peers. When code and data are not shared and methodology is not fully disclosed, peers do not have the ability to replicate the work and thus independent verification is impossible.*

A group of academics produced a paper on policy for releasing computer code, stating:

> *Not providing source code, they say, is now akin to withholding parts of the procedural process, which results in a "black box" approach to science, which is of course, not tolerated in virtually every other area of research in which results are published.*[204]

Even *The Guardian*, which has been a strong supporter of the IPCC activities and findings, urged:

> *If you're going to do good science, release the computer code too. Programs do more and more scientific work—but you need to be able to check them as well as the original data, as the recent row over climate change*

---

[204] http://wattsupwiththat.com/2012/04/17/the-journal-science-free-the-code/

documentation shows.[205]

Computer code is also at the heart of a scientific issue. One of the key features of science is deniability: if you erect a theory and someone produces evidence that it is wrong, then it falls. This is how science works: by openness, by publishing minute details of an experiment, some mathematical equations or a simulation; by doing this you embrace deniability. This does not seem to have happened in climate research. Many researchers have refused to release their computer programs—even though they are still in existence and not subject to commercial agreements.

Darrel Ince, the author of the article, is blunt in his conclusion:

So, if you are publishing research articles that use computer programs, if you want to claim that you are engaging in science, the programs are in your possession and you will not release them then I would not regard you as a scientist; I would also regard any papers based on the software as null and void.[206]

Leaked Emails Summary

A listing of the issues identified in the leaked emails provides the "what". What is more valuable at this point is to understand the "why". Many people have great difficulty accepting what they did, but even more difficulty in understanding why. These are scientists presumably working apolitically and objectively. The emails destroy all those illusions.

---

[205] http://www.guardian.co.uk/technology/2010/feb/05/science-climate-emails-code-release
[206] Ibid.

A few comments show they knew what they were doing was wrong:

Jones on February 21st, 2005:

> *I'm getting hassled by a couple of people to release the CRU station temperature data. Don't any of you three (authors of the hockey stick paper) tell anybody that the UK has Freedom of Information Act!*

Overpeck on June 28th, 2005:

> *Also, please note that in the US, the U.S. Congress is questioning whether it is ethical for IPCC authors to be using the IPCC to champion their own work/opinions. Obviously, this is wrong and scary, but if our goal is to get policymakers (liberal and conservative alike) to take our chapter seriously, it would only hurt our effort if we cite too many of our own papers (perception is often reality). PLEASE do not cite anything that is not absolutely needed, and please do not cite your papers unless they are absolutely needed. Common sense, but it isn't happening. Please be more critical with your citations so we have needed space, and also so we don't get perceived as self-serving or worse.*

Eystein Jansen on January 23rd, 2006:

> *Hi Peck [Overpeck], I assume the provisional acceptance is okay by IPCC rules? The timing of these matters are being followed closely by McIntyre and we cannot afford to being caught doing anything that is not within the regulations. Thus need to consult with Martin and Susan on this.*

Chapter Eleven

Jones on February 2nd, 2005:

Just send loads of station data to Scott. Make sure he documents everything better this time! And don't leave stuff lying around on FTP sites-you never know who is trawling them. The two MMs(McIntyre and McKitrick) have been after the CR eustachian data for years. If they ever hear there is a Freedom of Information act now in the UK, I think I'll delete the file rather than send to anyone. Does your similar act in the U.S. force you to respond to inquiries within 20 days,—our (sic) does! So the first request will test it. We also have a data protection act, the UK works on precedence, which I will hide behind. Tom Wigley has sent me a worried e-mail when he heard about it—thought people could ask him for his model code. He has retired officially from UEA so we can hide behind that. IPR (Intellectual Property Rights) should be relevant here, but I can see me getting into an argument with someone at UEA who say we must adhere to it!

Jones on February 21^{st,} 2005:

Even if WMO agrees, I will still not pass on the data. We have 25 or so years invested in the work. Why should I make the data available to you, when your aim is to try and find something wrong with it.

Jones on July 29th, 2005:

You can click on a reevaluation of MBH, which leads to a paper submitted to climatic change. This shows that in the H can be reproduced. The R-code to do this can be accessed and eventually the data—once the paper has

223

been accepted. IPCC will likely conclude that all and in arguments are wrong and have been answered in papers that have either come out or will soon." (He is talking about a report not yet produced, but with the insight of someone actively involved in the production.)

Jones July 5th, 2005:

I am reviewing a couple of papers on extremes, so that I can refer to them in the chapter for AR4. Somewhat circular, but I kept to my usual standards.

Jones July 5th, 2005:

If anything, I would like to see the climate change happen, so the science could be proved right, regardless of the consequences. This isn't being political it is being selfish.

Parker (UK Met office) February 2nd, 2005:

I have to say that I still back my initial reaction despite that being seen as a "diatribe". I included the salient emails as an appendix as there's some additional material in them that might help. At this stage I'm pretty sure we can reconcile these things relatively simply. However, I certainly would be unhappy to be associated with it if the current text remains through final draft—I'm absolutely positive it won't. As an aside for your eyes only (so please don't forward this part on to anyone) there may well be a very large signatory letter to BAMS from operational satellite guys that Fu et al. is wrong which is one reason why I want to avoid the impression given in zero order draft—it may make us all painted into a difficult corner.

Chapter Eleven

More will be obvious after TOVS workshop in May/June but things may move dramatically one way or the other so just a heads up. I note that my box on the lapse rates was completely and utterly ignored which may explain to some extent my reaction, but I also think the science is being manipulated to put a political spin on it which for all our sakes might not be too clever in the long run.

Jones on January 29th, 2004:

Mike, In an odd way this is cheering news! (commenting on death of John Daly owner web site "Still Waiting for Greenhouse") One other thing about the CC paper—just found another email—is that McKittrick says it is standard practice in Econometrics journals to give all the data and codes!! According to legal advice IPR (Intellectual Property Rights) overrides this.

Jones on November 16th, 1999:

I've just completed Mike's Nature trick of adding in the real temps to each series for the last 20 years (ie from 1981 onwards) amd [sic] from 1961 for Keith's to hide the decline.

Mann on July 31st, 2003:

p.s. I know I probably don't need to mention this, but just to insure absolutely clarity on this, I'm providing these for your own personal use, since you're a trusted colleague. So please don't pass this along to others without checking w/ me first. This is the sort of "dirty laundry" one doesn't want to fall into the hands of those who might potentially try to distort things...

There are numerous other examples, but these seem to destroy Phil Jones' or Mann's claim that this is regular banter between academic colleagues. That level of delusion is as hard to understand as the entire corruption of science. The best explanation for the group behavior is something called Groupthink.

The emails leaked from the Climatic Research Unit (CRU) in 2009 tell the story. People at the CRU were central to the Physical Science Basis Report of Working Group I of the IPCC and the Summary for Policymakers (SPM). The 2001 Report was most influential because it contained the "hockey stick"(HS). It was essential to protect it at all costs.

Groupthink

Irving Janis developed the concept of Groupthink, which enforces unanimity at the expense of quality decisions:

> Groups affected by groupthink ignore alternatives and tend to take irrational actions that dehumanize other groups. A group is especially vulnerable to groupthink when its members are similar in background, when the group is insulated from outside opinions, and when there are no clear rules for decision-making.[207]

The CRU/IPCC pattern appears to be a classic example.

Here's a list of some symptoms of groupthink with examples from CRU/IPCC emails and actions.

- **Having an illusion of invulnerability.** Content of

---

[207]

http://www.psysr.org/about/pubs_resources/groupthink%20overvie w.htm

the emails is a litany of arrogant invulnerability. Bradley's email is a good example. Notice the capital letter on Distinguished. In a backhanded way, Overpeck provides support for this position because he advised them on September 9th, 2009 to "Please write all emails as though they will be made public." Apparently he knew what they were saying was at least problematic, they didn't listen because they believed they were invulnerable. After the comments became public, they continued to believe they did nothing wrong. Michael Mann said:

> *What they've done is search through stolen personal emails—confidential between colleagues who often speak in a language they understand and is often foreign to the outside world...[turning]...something innocent into something nefarious.*[208]

CRU Director Phil Jones deflects by saying:

> *My colleagues and I accept that some of the published emails do not read well. I regret any upset or confusion caused as a result. Some were clearly written in the heat of the moment, others use colloquialisms frequently used between close colleagues...*

There is no confusion and it's arrogant and wrong to suggest others do it, as if that makes it acceptable.

- **Rationalizing poor decisions.** Jones rationalized the decision to withhold Freedom of Information (FOI) to the

---

[208]http://online.wsj.com/article/SB10001424052748703499404574559630382048494.html

University of East Anglia staff on December 3rd, 2008 as follows:

> Once they became aware of the types of people we were dealing with, everyone at UEA [in the registry and in the Environmental Sciences school—the head of school and a few others] became very supportive.

The entire body of emails supports this claim. Rob Wilson wrote on February 21st, 2006:

> I need to diplomatically word all this. I never wanted to criticise Mike's work in any way. It was for that reason that I made little mention to it initially.

On May 6^{th,} 1999, Mann wrote to Phil Jones:

> I trust that history will give us all proper credit for what we're doing here.

So do I!

Conversely, Keith Briffa battled with Mann and became increasingly alienated from the group. On June 17th, 2002 he wrote:

> I have just read this letter and I think it is crap. I am sick to death of Mann stating his reconstruction represents the tropical area just because it contains a few (poorly temperature representative) tropical series.

The PR battle involved proving superiority in credibility, rigour, and quality of published work.

- **Sharing stereotypes which guide the decision.** This

takes the form of unethical comments of practice going without challenge because they were all doing it. On September 19th, 1996 Funkhouser wrote:

> *I really wish I could be more positive about the Kyrgyzstan material, but I swear I pulled every trick out of my sleeve trying to milk something out of that.*

Mann said in a March 2003 email:

> *The Soon & Baliunas paper couldn't have cleared a 'legitimate' peer review process anywhere. That leaves only one possibility—that the peer-review process at Climate Research has been hijacked by a few skeptics on the editorial board.*

They ignore the fact that they were doing the same thing.

- **Exercising direct pressure on others.** On April 24th, 2003 Wigley wrote:

> *One approach is to go direct to the publishers and point out the fact that their journal is perceived as being a medium for disseminating misinformation under the guise of refereed work. I use the word 'perceived' here, since whether it is true or not is not what the publishers care about—it is how the journal is seen by the community that counts.*

In a direct threat, Phil Jones wrote:

> *I will be emailing the journal to tell them I'm having nothing more to do with it until they rid themselves of*

this troublesome editor.[209]

James Saiers, editor of *Geophysical Research Letters* was removed. A later email reported:

> *The GRL leak may have been plugged up now w/new editorial leadership there.*[210]

On October 14[th,] 2009, Trenberth expresses something to Tom Wigley that none of them ever dared say in public:

> *How come you do not agree with a statement that says we are nowhere close to knowing where energy is going or whether clouds are changing to make the planet brighter. We are not close to balancing the energy budget. The fact that we cannot account for what is happening in the climate system makes any consideration of geoengineering quite hopeless as we will never be able to tell if it is successful or not! It is a travesty!*

- **Maintaining an illusion of unanimity.**

Briffa struggles to maintain the illusion when he writes to Mann on April 29th, 2007,

> *I tried hard to balance the needs of the science and the IPCC, which were not always the same. I worried that*

209

http://www.ecowho.com/foia.php?file=1047388489.txt&search=peer+review
210

http://www.realclearpolitics.com/articles/2009/11/24/the_fix_is_in_99280.html

*you might think I gave the impression of not supporting
you well enough while trying to report on the issues and
uncertainties.*

On May 6th, 1999 Mann wrote to Phil Jones, "Trust that I'm
certainly on board w/you that we're all working towards a
common goal"

- **Using mindguards to protect the group from
  negative information.**

On December 10th, 2004, Schmidt gave the CRU gang a
Christmas present:

> *Colleagues, No doubt some of you share our frustration
> with the current state of media reporting on the climate
> change issue. Far too often we see agenda-driven
> "commentary" on the Internet and in the opinion columns
> of newspapers crowding out careful analysis. Many of us
> work hard on educating the public and journalists
> through lectures, interviews and letters to the editor, but
> this is often a thankless task. In order to be a little bit
> more pro-active, a group of us (see below) have recently
> got together to build a new 'climate blog' website:
> RealClimate.org which will be launched over the next few
> day...*
>
> *...The idea is that we working climate scientists
> should have a place where we can mount a rapid response
> to supposedly 'bombshell' papers that are doing the
> rounds and give more context to climate related stories or
> events.*

This was Mann's comment to the group about the establishment of RealClimate to act as "mindguards."

Some of the negative outcomes of groupthink also fit the actions of the CRU/IPCC group.

- **Examining few alternatives.** They narrowed the options by the definition of climate change to only those caused by human activities. Of the three greenhouse gases, almost all the focus is on CO_2.

- **Not being critical of each other's ideas.** Not only were they not critical, but they peer reviewed each other's work and controlled who they recommended to editors as reviewers. Mann to Jones on June 4th, 2003:

    I'd like to tentatively propose to pass this along to Phil as the "official keeper" of the draft to finalize and submit IF it isn't in satisfactory shape by the time I have to leave.

On August 5th, 2009 Jones wrote to Grant Foster in response to his request for reviewers for an article:

    I'd go for one of Tom Peterson or Dave Easterling. To get a spread, I'd go with 3 US, One Australian and one in Europe. So Neville Nicholls and David Parker. All of them know the sorts of things to say—about our comment and the awful original, without any prompting.

- **Not examining early alternatives.** There was a graph of temperatures drawn by Lamb showing the Medieval Warm Period (MWP) and used in the first IPCC Report. It was correct, but contradicted their claim of modern warming. As Mann said to Jones on June 4th, 2003:

...it would be nice to try to "contain" the putative "MWP", even if we don't yet have a hemispheric mean reconstruction available that far back.

They chose to rewrite history.

- **Not seeking expert opinion.** Professor Wegman spoke directly to this problem in his report for the U.S. Congress on the infamous hockey stick graph.

  It is important to note the isolation of the paleoclimate community; even though they rely heavily on statistical methods they do not seem to be interacting with the statistical community.[211]

- **Being highly selective in gathering information.** Apart from only looking at human causes, the CRU emails have many examples of data selected to prove their point. Tim Osborn to the group on October 5th, 1999 speaks of the issue McIntyre identified of truncated records:

  They go from 1402 to 1995, although we usually stop the series in 1960 because of the recent non-temperature signal that is superimposed on the tree-ring data that we use.

On the March 19th, 2009, Santer wrote to Jones about the Royal Meteorological Society (RMS) asking for data used for a publication:

If the RMS is going to require authors to make ALL data

---

[211]http://www.uoguelph.ca/~rmckitri/research/WegmanReport.pdf

available—raw data PLUS results from all intermediate calculations—I will not submit any further papers to RMS journals.

On September 27th, 2009, Tom Wigley wrote to Phil Jones about a problem with Sea Surface Temperatures (SST):

So, if we could reduce the ocean blip by, say, 0.15 deg C, then this would be significant for the global mean— but we'd still have to explain the land blip.

- **Not having contingency plans.** They never dreamed they would be exposed. Part of the lack of contingency plans may be attributed to the fact they had the backing of powerful people and involvement of public relations experts to block, offset and counterattack. The Chapter 8 incident occurred early in the process of the IPCC and undoubtedly gave them confidence they were protected.

It didn't work because of poor judgment. Janis explains in groupthink it "occurs when a group makes faulty decisions because group pressures lead to a deterioration of "mental efficiency, reality testing, and moral judgment." What went on at the CRU and the IPCC appears to be a classic example.

Now they hide behind the fact that most can't believe scientists could ignore or deliberately manipulate data, distort procedures and not have more of them speak out. They also can't believe a small group of people could achieve such deception.

GAVIN SCHMIDT

CHAPTER TWELVE

How Did So Few Achieve Such a Large Deception?

Specialized meaninglessness has come to be regarded, in certain circles, as a kind of hallmark of true science.
—Aldous Huxley, (1894-1963) British Author

The thing from which the world suffers just now more than any other evil is not the assertion of falsehood, but the endless repetition of half-truths.
—G. K. Chesterton (1874-1936) British Author

TO PARAPHRASE WINSTON Churchill; never have so many been deceived by so few, at so great a cost.

On May 6th, 1999, Michael Mann wrote to Phil Jones:

Trust that I'm certainly on board w/ you that we're all working towards a common goal. That is what is distressing about commentarys [sic] (yours from last year, and potentially, without us having had appropriate [sic] input, Keith and Tim's now) that appear to "divide and conquer". The skeptics happily took your commentary last year as reason to doubt our results! In fact, your piece was references [sic] in several commentaries (mostly on the WEB, not published) attacking our work. So THAT is what this is all about. It is in the NAME of the common effort we're all engaged in, that I have voiced concerns about language and details in this latest commentary— so as to avoid precisely that scenario.

This is a rambling discourse that is a disturbing insight into what was going on at the CRU. What is the common goal he mentions? The necessity to isolate CO_2 and prove scientifically it was destroying the planet. Which explains his concern about what he perceives as negative and divisive comments. The difficulty is: it is essential to question, challenge and investigate to achieve proper science. Mann is saying that by identifying the problems and limitations, they are giving comfort to the enemy. This further confirms the political nature of activities at the CRU and the IPCC. The connection is easy to make because Mann and Jones were the main authors of the critical pieces of evidence of human causes of warming in the 2001 IPCC Report.

This raises questions about how and why so many people, especially scientists, were taken in by the deception that concerns Mann. It is instructive to understand how different groups became part of the so-called consensus. The core IPCC people were carefully selected and most of them worked at the CRU. The *Ad Hoc Committee Report on the 'Hockey Stick' Global Climate Reconstruction* commonly known as *The Wegman Report* said:

As analyzed in our social network, there is a tightly knit group of individuals who passionately believe in their thesis. However, our perception is that this group has a self-reinforcing feedback mechanism and, moreover, the work has been sufficiently politicized that they can

236

hardly reassess their public positions without losing credibility.

Wegman identified most of the people involved with the leaked information from the CRU.

IPCC participants are chosen by the national weather agencies of each member of the World Meteorological Organization (WMO). The Intergovernmental Panel on Climate Change (IPCC) required people who would achieve the political and scientific objective of identifying human activities as the cause of global warming, and later climate change, generally referred to as Anthropogenic Global Warming (AGW) theory. Their work effectively thwarted the standard scientific method of disproving the theory. Scientists who dared to question the theory were derisively called skeptics. When this epithet didn't stop them, they were called deniers with its holocaust connotations. Most of the so-called skeptics were well qualified but excluded from the IPCC, making it a carefully selected group.

Some, such as Richard Lindzen[212], Alfred P Sloan professor of meteorology at MIT, participated—hoping to have reasonable scientific input but eventually gave up. *There's little doubt*, Lindzen said, *that the IPCC process has become politicized to the point of uselessness.*

How did the IPCC maintain control and convince many, including political leaders, they were right and were the authority? Beyond using UN agencies as vehicles, they had the challenge of running an apparently open process while keeping total control.

They controlled who participated and who were the lead authors, especially of critical chapters. Richard Lindzen notes:

---

[212] http://news.heartland.org/newspaper-article/2001/06/01/ipcc-report-criticized-one-its-lead-authors

> *IPCC's emphasis, however, isn't on getting qualified scientists, but on getting representatives from over 100 countries...the truth is only a handful of countries do quality climate research. Most of the so-called experts served merely to pad the numbers.*

They published the political document, the Summary for Policymakers (SPM) before the Technical (Science) Report of Working Group I was issued. Making sure the Technical Report matched the SPM. Lindzen again: *The IPCC clearly uses the Summary for Policymakers to misrepresent what is in the report.*

They used wording in the SPM to catch the media and public attention. It's difficult to describe scientific information for an essentially non-scientific audience through the media; what one blogger describes as the "Math-Free Zone of Journalism". Columnist James Kilpatrick says, "People who write for a living should never be left alone with mathematics. They are almost bound to mess up." They are less likely to with the terms created by the IPCC, but it is easier to dramatize. Using non-mathematic terminology in the SPM, exemplified by the labels set out in a table in the third report, such as; Very unlikely (1–10 %) Likely (66–90 %) Very likely (90–99 %). The percentages are not used in the Technical Report. As one study says, "How the assessment frames the information is determined by the choices and goals of the users." For the IPCC, this includes focusing on negative impacts of warming when there are positive effects and including and highlighting studies that appeared to identify a "human signal".

Here is Lindzen's summary of the IPCC process:

> *It uses summaries to misrepresent what scientists say; uses language that means different things to scientists and laymen; exploits public ignorance over quantitative matters; exploits what scientists can agree on while ignoring disagreements to support the global warming*

*agenda; and exaggerates scientific accuracy and certainty
and the authority of undistinguished scientists.*

The rest of this chapter examines some of the issues Lindzen identifies, events that explain how they achieved the global deception.

A major reason the deception was easy involved public lack of knowledge of the extent and nature of climate change. They have no idea how temperatures are reconstructed, illustrated by the common question I get at a presentation; "How do they know what the temperature was years ago?" There are three segments of temperature data. The boundaries between temperature periods are the result of technical or methodological measures not natural factors. The instrumental record that covers just over 100 years, the historical record of human observations covering about 3,000 years and the rest of time is in the biologic or geologic time. Most records, especially in the historical period, are called proxy records, that is, they are a secondary indication of weather and climate such as the first arrival of geese in the spring or the date of cherry blossoms in Japan.

All records contradict claims made by official climate science. They show:

- Temperatures vary considerably and in very short time periods.
- Global temperatures were much warmer than today on many occasions.
- Temperature increases precede CO_2 increases.
- Current changes are not unprecedented.

The Antarctic ice core record covering 420,000 years appeared in 1991. (see Figure 30) It shows how much temperatures varies with a range of some 12°C. This became sidelined by the fact they also published the varying CO_2 over the same time period.

File: Vostok Petit data.svg
From Wikipedia, the free encyclopedia

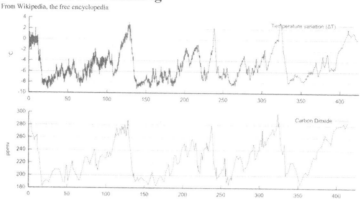

Figure 30

Antarctic ice core record. Temperature top and CO_2 bottom

The relationship was immediately seized and promoted as evidence that CO_2 was driving temperature. Only a few years later we learned that the relationship is exactly the opposite. We now know that in every record of any duration for any period, temperature increases before CO_2. This completely contradicts and therefore negates the basic assumption of the AGW hypothesis.

Despite limitations, such as a 70-year smoothing average, ice core records provide another contradiction to a major IPCC claim, namely that the world is warmer than ever before.

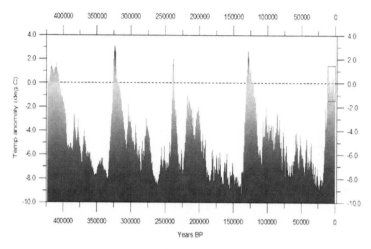

Figure 31

Antarctic temperature record showing previous Interglacial peaks.[213]

The graph in Figure 31 shows how much temperature changes naturally with dramatic swings over a range of 12°C. These swings are greater when you understand that they applied 70 year smoothing average. This equates to using a single temperature for the last 70 years. Which 70 years would you choose as 'better' than today? Think of recent cold winters and consider whether it was better. The world was almost as warm as today for just a fraction of the last half million years. Cold is the predominant condition.

Despite that, the world was warmer than today for most of the last 10,000 years: a period variously known as the Climatic Optimum, the Hypsithermal and now the Holocene Optimum. It is recorded in the Greenland ice core.

---

[213] http://climate4you.com/

Greenland GISP2 Ice Core - Temperature Last 10,000 Years

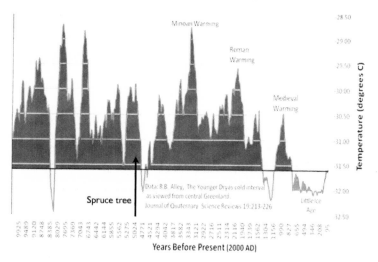

Figure 32

Greenland Temperature Graph—World Warmer than Today for Most of the Last 10,000 Years

It was 3°C warmer on average than today for most of the 10,000 years. Tangible evidence of warmth is a fossilized white spruce (Picea glauca) photographed by Professor Ritchie (Figure 33) and radiocarbon dated at 4940 ±140 years Before Present (BP) as indicated on the ice core chart.

Figure 33

**5,000 Year Old White Spruce Stump North of Current
Tree Line**

Note the existence of the Medieval Warm Period (MWP) on the
right side of the graph. This is what they had to eliminate as
discussed earlier. More important, note the world was warmer
than today for most of the preceding 8000 years. This gives lie to
claims about demise of the polar bear because they survived these
conditions.

Some records are valuable because they transcend the
boundaries. Tree ring records are a good example, but this made
them ideal for exploitation of climate science. It worked for the
climate deception because few, including most scientists, know
how inadequate they are for accurate reconstruction or how easily
they're distorted. Tree rings are a source of data used to span the
instrumental and history stages of reconstruction. The original
use, called dendrochronology, determined age of the tree by
counting annual growth rings. Then, with the work of people like
A.E. Douglass, they became valuable in reconstructing solar

activity as they registered variation carbon 14 in the atmosphere. The use for the IPCC record was completely inappropriate.

They used tree rings for climate studies with the assumption that they indicate temperature change. In reality, they reflect the growth pattern and are the result of a multitude of environmental factors. Temperature is a minor factor. Precipitation is the main determinate of growth, as any gardener knows.

Despite this a few, especially those associated with the Climatic Research Unit (CRU) of East Anglia, began producing studies using tree rings solely as an indicator of temperature. The pivotal paper published in Nature in 1998 by Mann, Bradley and Hughes titled, *Global-scale temperature patterns and climate forcing over the past six centuries* become known as MBH98. Mann was the principal author, so John Daly gave him credit and wrote "Mann completely redrew the history, turning the Medieval Warm Period and Little Ice Age into non-events, consigned to a kind of Orwellian 'memory hole'". The tree ring data he produced formed the handle of the stick. Then, using an inappropriate technique he tacked on the modern instrumental record to form the blade. We later learned this was necessary because the tree ring data showed declining temperatures for the 20th century. It was terrible science and statistically wrong. Despite this, the paper passed peer review.

The hockey stick graph, remarkably quickly, became the orthodoxy. As John Daly explained:

> What is disquieting about the 'Hockey Stick' is not Mann's presentation of it originally. As with any paper, it would sink into oblivion if found to be flawed in any way. Rather it was the reaction of the greenhouse industry to it—the chorus of approval, the complete lack of critical evaluation of the theory, the blind acceptance of evidence which was so flimsy. The industry embraced

*the theory for one reason and one reason only—it told them exactly what they **wanted** to hear.*[214] *[emphasis in original]*

Mann became lead author of the chapter *Observed Climate Variability and Change* in the 2001 IPCC Report. He was also a contributing author on other chapters. The hockey stick received prominence in the 2001 IPCC Report. Ross McKitrick wrote:

It was central to the 2001 Third Assessment Report [TAR] from the Intergovernmental Panel on Climate Change (IPCC). It appears as Figure 1b in the Working Group 1 Summary for Policymakers, Figure 5 in the Technical Summary, twice in Chapter 2 (Figures 2-20 and 2-21) of the main report, and Figures 2-3 and 9-1B in the Synthesis Report. Referring to this figure, the IPCC Summary for Policymakers (p.3) claimed it is likely "that the 1990s has been the warmest decade and 1998 the warmest year of the millennium" for the northern hemisphere.[215]

He also notes the importance of the hockey stick to their scientific agenda, designed to support the political agenda, measured by the different highlighting from other information in the report:

In appreciating the promotional aspect of this graph, observe not only the number of times it appears, but its size and colorful prominence every time it is shown.[216]

---

[214] http://www.john-daly.com/hockey/hockey.htm

[215] http://www.uoguelph.ca/~rmckitri/research/McKitrick-hockeystick.pdf

[216] Ibid.

The control of what went into the Technical Summary (Science Report) and the SPM by just a few people is the real issue and critical to understanding how a few people controlled the deception that fooled the world.

Mann's work provided the handle for the hockey stick. He rewrote history by eliminating the MWP, but the hockey stick has a blade with data provided effectively by Phil Jones, Director of the CRU and IPCC lead author. His work claimed temperatures after the LIA rose at a rate greater than any in the natural record and thus indicated a human signal. In the SPM, the hockey stick and temperature graphs appear together as the bottom figure of Figure 34.

Variations of the Earth's surface temperature for:

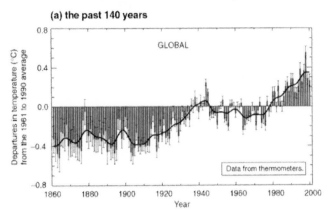

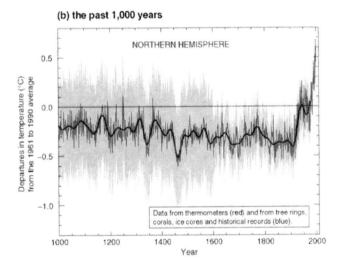

Figure 1: Variations of the Earth's surface temperature over the last 140 years and the last millennium.

Figure 34

Variations of the Earth's Surface Temperature

Jones claimed an increase of 0.6°C in the global average annual temperature in approximately 130 years. The actual statement in the SPM is odd:

> *Over both the last 140 years and 100 years, the best estimate is that the global average surface temperature has increased by 0.6 ± 0.2°C.*[217]

IPCC claim this increase is beyond any natural increase and, therefore, anthropogenic.

This is simply incorrect. Actually, it is within the error factor of calculations of global average temperatures. Besides, there are so many problems with the data many consider it impossible to calculate the global temperature. The error range of ±0.2°C, which is a ± 33 percent, shows the problem. The IPCC Report identifies some of these:

> *There are uncertainties in the annual data (thin black whisker bars represent the 95% confidence range) due to data gaps, random instrumental errors and uncertainties, uncertainties in bias corrections in the ocean surface temperature data and also in adjustments for urbanisation over the land.*[218]

Here are some of the other problems:

- There are very few records, approximately 1000, of 100 years length or more, as this Goddard Institute of Space Studies Plot (GISS) shows (Figure 35).

---

[217] http://www.grida.no/publications/other/ipcc_tar/
[218] http://www.grida.no/publications/other/ipcc_tar/

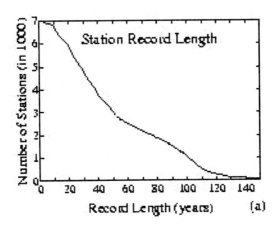

Figure 35

Station Record Length

- Most are concentrated in eastern North America and Western Europe.
- Most of these stations are affected by the Urban Heat Island effect that artificially increases the temperature.
- Instruments varied regionally and over time, but because of early limitations all records only measure to $0.5°$
- There are virtually no measurements for the oceans, which are 70% of the surface.
- Few measurements exist for the deserts (19% of the land surface), mountains (20%), or forest 40%.

There is serious scientific concern about the nature, length and quality of the data base best expressed by the U.S. National Research Council Report in 1999:

> *Deficiencies in the accuracy, quality and continuity of the records place serious limitations on the confidence that can be placed in the research results.*

The actual figures Jones gave, which were 0.6°C ± 0.2°C, an error factor of ± 33% underscore this problem. The limitations and the error factor are sufficient to reject the argument that it represents a real increase. It is completely inadequate to serve as part of the basis for global climate and energy policies.

But, there is a more serious problem. We are unable to reproduce Jones' results because he refused to disclose which stations he used and how he adjusted the data. On February 21st, 2005 in response to a request from Warwick Hughes, an Australian researcher who has long sought to verify the global temperature record, Jones wrote:

> We have 25 or so years invested in the work. Why should I make the data available to you, when your aim is to try and find something wrong with it.

Apparently Jones is not alone in the practice of non-disclosure or denial of access to climate data. Finally, we learned from Jones that the original data was lost. At least he acknowledged this was unacceptable.[219]

A series of attempts to obtain information from the University of East Anglia and from the joint enterprise of the Hadley Centre and the Climate Research Unit known as HadCRUT3 appear on the ClimateAudit Blog site. This site also discusses disturbing questions about modifications to past records apparently to make the 1930s appear cooler, thus enhancing the claim that the world is warmer than it has ever been.

The most recent 'human signal' is not actual evidence at all. It comes from carefully manipulated computer models designed to isolate a portion of temperature increase as clearly human. People are generally unaware that all 'predictions' of global warming

---

[219] http://www.theguardian.com/environment/2010/feb/15/phil-jones-lost-weather-data

come from computer models. These can't work because the database that limits Jones is totally inadequate for models. Another factor compounds the problem. While Jones estimated surface temperature, the models are three-dimensional and we have virtually no data for the atmosphere. Combine this with the limited knowledge of atmospheric, oceanic, solar mechanisms necessary for a model to work and it is no surprise the models fail to simulate past climates or accurately predict future ones. Models that can't forecast weather beyond 7 days are incapable of predicting conditions 30, 40 or 100 years from now.

It is impossible to address all the errors in the science, assumptions, methods, data, and computer models. So far I have examined the objective, motive and some of the major gaming carried out. However, it is all ultimately tested in the real world. A simple but powerful definition of science is the ability to predict. If your predictions are wrong there is clearly something wrong with your science. Weather forecast failures indicate it is not a science. Supporters of "official" climate science, produced by the IPCC, tried to distance themselves from this problem by saying that they were two different things. The difficulty is climate is an average of the weather; therefore it can only be as precise as the weather.

Every single climate prediction (projection) the IPCC made has been wrong. As we saw earlier, they ostensibly switched from predictions to projections because of the failures. They then produced three scenarios based upon economic development that would determine the amount of human CO_2 produced. For a while it appeared that temperature was increasing approximately in line with CO_2 increase. It is likely that much of this was due to manipulation of the major components including the temperature and human production of CO_2.

Another Reason Why IPCC Predictions (Projections) Fail. AR5 Continues to Let The End Justify the Unscrupulous Means

Someone said economists try to predict the tide by measuring one wave. The IPCC essentially try to predict (project) the global temperature by measuring one variable. The IPCC compound their problems by projecting the temperature variable with the influence of the economic variable.

Use of circular arguments is standard operating procedure for the IPCC. For example, they assume a CO_2 increase causes a temperature increase. They then create a model with that assumption and when the model output shows a temperature increase with a CO_2 increase they claim it proves their assumption.

They double down on this by combining an economic model that projects a CO_2 increase with their climate model projection. To make it look more accurate and reasonable they create scenarios based on their estimates of future developments. It creates what they want, namely that CO_2 will increase and temperature will increase catastrophically unless we shut down fossil fuel based economies very quickly.

All their projections failed, even the lowest as, according to them, atmospheric CO_2 continued to rise and global temperatures declined. As usual, instead of admitting their work and assumptions were wrong, they scramble to blur, obfuscate and counterattack.

One part of the obfuscation is to keep the focus on climate science. Most think the IPCC is purely about climate science, they don't know about the economics connection. They don't know that the IPCC projects CO_2 increase on economic models that presume to know the future. Chances of knowing that are virtually zero, as history shows.

On September 1st, 2014 we will recognize the 75th anniversary of the declaration of war against Germany. I am not aware of anybody who predicted what happened in that 75 years, or even came close. I am sure people will find someone who foresaw one or two of the events, but not the entire social, economic, technological and political changes. A brief list illustrates the challenge:

- The Cold War
- The Korean War
- The Vietnamese War
- Global Terrorism
- The collapse of communism
- China and India as world powers
- The Internet
- Moon and Mars Landings
- Silicon Chips
- Space vehicle leaving the Solar System
- Space Satellites
- Hubble telescope
- Fracking

The IPCC claim 95 percent certainty about their climate science and presumably about their predictions. The problem is all were wrong from the start. As early as the 1995 Report they had switched to projections. They gave a range of projections or scenarios from low to high, but even the lowest was incorrect. Roger Pielke Jr., explained the assumptions for the scenarios were unrealistic, especially about technological progress in energy use and supply.[220]

Most people assume the projections are solely a function of the climate science and climate models, but that is not the case. The climate science is wrong and that contributes to the failed projections because it is the basic assumption of the AGW hypothesis that an increase in CO_2 causes a temperature increase. However, the three projections also vary from high to low because of different assumptions about the future society and economy. These estimates of the future primarily determine the amount of CO_2 increase that will occur under different economic scenarios. As Richard Lindzen, MIT professor of meteorology said in an interview with James Glassman that the 2001 IPCC Report "was very much a children's exercise of what might possibly happen" prepared by a "peculiar group" with "no technical competence." Maybe, but it achieved their political objective of isolating and demonizing CO_2.

After release of the 2001 Third Assessment Report (TAR) two papers by Ian Castles and David Henderson (C&H) were published drawing attention to the problems with the emission scenarios used to produce the three projections.[221] Castles explained the concerns as follows:

> *During the past three years I and a co-author (David Henderson, former Head of the Department of Economics and Statistics at OECD) have criticised the IPCC's treatment of economic issues.*

[220] http://rogerpielkejr.blogspot.ca/2010/05/misrepresentation-of-ipcc-co2-emission.html
[221] Ian Castles and David Henderson (2003) Economics, emissions scenarios and the work of the IPCC, *Energy & Environment*, vol. 14, no. 4 and Ian Castles and David Henderson (2003) The IPCC emission scenarios: An economic-statistical critique, *Energy & Environment*, vol. 14: nos.2-3.

Chapter Twelve

Our main single criticism has been the Panel's use of exchange rate converters to put the GDPs of different countries onto a common basis for purposes of estimating and projecting output, income, energy intensity, etc. This is not permissible under the internationally-agreed System of National Accounts which was unanimously approved by the UN Statistical Commission in 1993, and published later that year by the United Nations, the World Bank, the IMF, the OECD and the Commission of the European Communities, under cover of a Foreword which was personally signed by the Heads of the five organisations. [222]

As one commentator noted:

These two economists have shown that the calculations carried out by the IPCC concerning per capita income, economic growth and greenhouse gas emissions in different regions are fundamentally flawed, and substantially overstate the likely growth in developing countries. The results are therefore unsuitable as a starting point for the next IPCC assessment report, which is due to be published in 2007. Unfortunately, this is precisely how the IPCC now intends to use it submissions projections. [223]

The IPCC response was typical of the arrogant superiority and belief in their unassailability that pervades most of their dealings:

---

[222] http://climateaudit.org/2005/08/22/ian-castles-on-ipcc-economic-assumptions/
[223] http://www.lavoisier.com.au/articles/climate-policy/economics/castles-hendersonresponse.pdf

> *On December 8th, 2003 at the Milan COP9 Dr. Pachauri*
> *released a press statement which criticized the arguments*
> *which Castles in Henderson have been making in this*
> *debate.*

Pachauri's charges against C&H, especially Castle's were false personal attacks.

Richard Tol commented on C&H and the IPCC response:

> *Castles and Henderson....criticized the IPCC for using*
> *market exchange rates in the economic accounting used as*
> *a basis for its SRES scenarios. This started as a technical*
> *dispute. However, the initial IPCC response—which can*
> *be characterized as "We are the IPCC. We do not make*
> *mistakes. Please go away."—raised the stakes and turned*
> *the debate into one about the credibility of the entire*
> *IPCC, a debate that now includes politicians and the*
> *public. Howard Herzog of MIT recently summarized this*
> *as the "IPCC is a four letter word".*

The UNFCCC predetermined the results of the IPCC work by directing with them to study only human causes of climate change. The IPCC then narrowed the focus to human produced CO_2 as the cause of warming. They directed their efforts to proving rather than disproving their hypothesis. Central to this objective was the need to have atmospheric CO_2 levels rise constantly because of a constant rise in human production of CO_2.

The IPCC controlled results of rising atmospheric levels with data from warming advocate Charles Keeling's, and later his son Roger's, measurements at Mauna Loa. There is fascinating, but disturbing correspondence on this issue between Ernst Georg Beck and Roger Keeling. Beck had to be dismissed because his work showed that 19th century levels of atmospheric CO_2 were much higher than used by the IPCC and created by Guy

Callendar and Tom Wigley. The IPCC controlled the production of the annual increase in human production of CO_2.

In their 2001 Report the IPCC note the increase of CO_2 from 6.5 GtC (gigatons of carbon) human sources to 7.5 GtC in the 2007 report. In the FAQ section they answer the question "How does the IPCC produce its Inventory Guidelines?" as follows:

> *Utilizing IPCC procedures, nominated experts from around the world draft the reports that are then extensively reviewed twice before approval by the IPCC.*[224]

In a 2008 article Castles notes about the 2007 Report:

> *Unfortunately, the assumptions it uses overstate potential manmade global warming by a large measure.*
>
> *In 2001 IPCC based its predictions of substantially warming temperatures during the next century largely on forecasts of explosive growth in Third World economies—and hence emissions—during the twenty-first century. The panel actually predicted Third World nations would grow so fast they would surpass the economies of wealthy Western nations.*[225]
>
> *Economists pointed out the unrealistic assumptions, but in the six years since these IPCC gaffes, little appears to have changed.*

Richard Tol commented on the changes for AR5:

---

[224] http://www.ipcc-nggip.iges.or.jp/faq/faq.html
[225] http://news.heartland.org/newspaper-article/2008/03/01/economic-formulas-ipcc-report-criticized-overstating-emissions

IPCC AR5 of Working Group 1 will therefore be based on scenarios-formerly-known-as-SRES. They're now called RCP.[226]

A presentation Representative Concentration Pathways (RCPs) by Jean-Pascal van Ypersele, Vice Chair of the IPCC, lays out the challenge.[227]

Some of the Challenges for AR5

- **Improve policy-relevance, without becoming policy-prescriptive**
- **Improve quality and readability**
- **Provide elements of answer to difficult/new questions (+ some treated as FAQ)**
- **Integrate Synthesis Report « design » in the scoping process from the start**
- **Improve developing countries participation**

Jean-Pascal van Ypersele
(vanypersele@astr.ucl.ac.be)

In a classic bureaucrat flow chart he shows a change in process that

---

[226] https://tallbloke.wordpress.com/2013/03/14/sres-scenarios-that-shaped-energy-policy-for-our-economic-future-had-greenpeace-input/
[227] http://unfccc2.meta-fusion.com/kongresse/090601_SB30_Bonn/downl/20090603_Yperse le.pdf

among other things appears to make the role of economic development unclear.

New scenarios development process – parallel vs. sequential approach

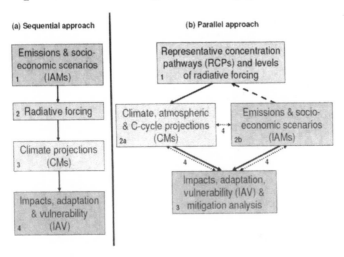

William Kininmonth, former head of Australia's National Climate Centre and their delegate to the WMO Commission for Climatology and author of the insightful book *Climate Change: A Natural Hazard* wrote the following in an email on the *ClimateSceptics* group page:

> *I was at first confused to see the RCP concept emerge in AR5. I have come to the conclusion that RCP is no more than a sleight of hand to confuse readers and hide absurdities in the previous approach.*
>
> *You will recall that the previous carbon emission scenarios were supposed to be based on solid economic models. However this basis was challenged by reputable*

economists and the IPCC economic modeling was left rather ragged and a huge question mark hanging over it.

I sense the RCP approach is to bypass the fraught economic modeling: prescribed radiation forcing pathways are fed into the climate models to give future temperature rise—if the radiation forcing plateaus at 8.5W/m2 sometime after 2100 then the global temperature rise will be 3C. But what does 8.5 W/m2 mean? Previously it was suggested that a doubling of CO_2 would give a radiation forcing of 3.7 W/m2. To reach a radiation forcing of 7.4 W/m2 would thus require a doubling again—4 times CO_2 concentration. Thus to follow RCP8.5 it is necessary for the atmospheric CO_2 concentration equivalent to exceed 1120ppm after 2100.

We are left questioning the realism of a RCP 8.5 scenario. Is there any likelihood of the atmospheric CO_2 reaching about 1120 ppm by 2100? IPCC has raised a straw man scenario to give a 'dangerous' global temperature rise of about 3C early in the 22nd century knowing full well that such a concentration has an extremely low probability of being achieved. But, of course, this is not explained to the politicians and policymakers. They are told of the dangerous outcome if the RCP8.5 is followed without being told of the low probability of it occurring.

One absurdity is replaced by another! Or have I missed something fundamental?[228]

I don't think he has. In reality, it doesn't matter whether it changes anything because the underpinning of the climate science and the economics depends on accurate data and knowledge of mechanisms.

---

[228] Reproduced with permission of William Kininmonth.

Chapter Twelve

We know there was insufficient weather data on which to construct climate models and the situation deteriorated as they eliminated weather stations and 'adjusted' then cherry-picked data. We know knowledge of mechanisms is inadequate because the IPCC WGI Science Report says so:

> *Unfortunately, the total surface heat and water fluxes (see Supplementary Material, Figure S8.14) are not well observed.*

or:

> *For models to simulate accurately the seasonally varying pattern of precipitation, they must correctly simulate a number of processes (e.g., evapotranspiration, condensation, transport) that are difficult to evaluate at a global scale.*

In a perverse way the IPCC acknowledge this with their attempt to claim the "pause" in temperatures of the last 15 years was due to some "deep ocean" process. Again Kininmonth acutely observes the comment in the SPM that:

> *There may also be a contribution from forcing inadequacies and, **in some models, an overestimate of the response to increasing greenhouse gas and other anthropogenic forcing** (dominated by the effects of aerosols)." [My emphasis]*

> *With the inability to explain with confidence the 15 year temperature pause this is rather damning. (Two potential explanations are given for the pause, one with low confidence and the other with only medium confidence— i.e., guesswork.) It is difficult for the acolytes to now*

shout us down with "The science is settled!"

Economic projections are even more difficult because of lack of data, an inability to anticipate public feedback and political reaction, but primarily the impossibility of anticipating technology and innovation. That is the critical part of the list of events in the last 75 years that completely changed the direction of history. It guaranteed that any predictions or projections would be wrong—the IPCC projections will be wrong for the same reason, but with the added problem of bad science. They must know this, so it only underscores the political nature of their work.

They've already shown that being wrong or being caught doesn't matter because the objective of the scary headline is achieved by the complete disconnect between their Science Reports and the Summary for Policymakers. It is also no coincidence that the SPM is released before national politicians meet to set their budgets for climate change and the IPCC. As Saul Alinsky insisted in rules for radicals the end justifies the means.[229]

---

[229] http://www.crossroad.to/Quotes/communism/alinsky.htm

ALBERT ARNOLD GORE, JR

CHAPTER THIRTEEN

What's Next?

The Big Lie

All this was inspired by the principle—which is quite true in itself—that in the big lie there is always a certain force of credibility; because the broad masses of a nation are always more easily corrupted in the deeper strata of their emotional nature than consciously or voluntarily.
—Adolf Hitler, Mein Kampf

The Precautionary Principle Excuse

In a free society the individual is presumed to be free to act unless the state can prove harm or the potential to do harm. The precautionary principle says that no individual person is free to act unless that individual can prove to the state that the action can do no harm.
—Aaron Wildavsky

So, as a political phenomenon, the "principle" is a

logically unsupported attempt to justify totalitarianism?
—Kesten Green

Public Relations Deception

Want good coverage? Tell a good story. When your business is under siege, you can't hope to control the situation without first controlling the story. The most effective form of communication is a compelling narrative that ties your interest to those of your audience. This is particularly critical when you're caught in the spotlight; it doesn't matter if you have the facts on your side if your detractors are framing the story. So, don't just react. Take some time now to define your company story. Then you'll be ready to build a response into that narrative should something go wrong.
—Jim Hoggan, president of Hoggan and Associates in the Vancouver Sun December 30th, 2005.

JIM HOGGAN IS the owner of a leading Canadian PR company. He is Chairman of the Board of the David Suzuki Foundation, a major Canadian environmental group. He is also founder and sponsor of DeSmogBlog, a web site dedicated to attacking anyone who dares to ask questions about climate or environmental science. The web site only exists to promote a political agenda and attack anyone who dares to point out the errors in the science. They apparently agree with the view of Greenpeace co-founder Patrick Watson who said, "It doesn't matter what is true, it only matters what people believe is true." Read Hoggan's comment above and you can see a similar mentality. The trouble is environment and climate issues are about science not spin.

The PR approach pervades the entire exploitation of climate and environment. Every time a serious problem cropped up for

IPCC official climate science—or those promoting it—they hired professional spin-doctors. Why do 'official' climate scientists need spin-doctors? Answer: because they practice politics, not science. Climategate exposed the disgraceful behavior at the CRU as they worked to create the science required by Maurice Strong and the people orchestrating political control by pretending to save the planet. They hired PR experts to direct diversions, deceptions and cover-ups.

The greatest deception in history used science and the new paradigm of environmentalism to demonize CO_2. This was done to achieve the Club of Rome goal verbalized by Maurice Strong to shut down the industrialized nations. They determined that resolution of a global problem needs a one-world government. A skillfully orchestrated campaign by a cabal of powerful people, evolved from the Club of Rome, and implemented their objective through the UN. As Elaine Dewar explained:

> *Strong was using the U.N. as a platform to sell a global environment crisis and the Global Governance Agenda.*

He created the United Nations Environment Program (UNEP) within which he created the necessary political and scientific apparatus. There is nothing wrong or illegal with a political objective. The wrongdoing and illegalities are in how it was achieved. The ongoing problems are in the damage even partial implementation of the programs has created.

The world learned about the plan at Rio de Janeiro in June 1992.[230] The principle theme was the environment and sustainable development. The documents that resulted were Agenda 21, the Rio Declaration on Environment and Development, the Statement of Forest Principles, and the United Nations Framework Convention on Climate Change (UNFCC) and the

---

[230] http://www.un.org/geninfo/bp/enviro.html

United Nations Convention on Biological Diversity. The UNFCC produced the definition of climate change that the Intergovernmental Panel on Climate Change (IPCC) used to narrow research to only human causes of climate change. This allowed them to produce the scientific condemnation of CO_2, the byproduct of the Industrial Revolution that was destroying the planet.

They used science, but if it was inadequate they invoked the precautionary principle, enshrined in Principle 15 of Agenda 21. It is the standard environmentalists fallback position; we should act anyway, just in case. A global climate threat was necessary to achieve the goal of a world government. Working through the UN had several advantages for their goal as Maurice Strong said, they could get all the money they wanted and not be accountable to anyone.

From all reports, Strong is a master organizer and knows the powers of bureaucracies. The UN gave access to the bureaucracy of the World Meteorological Organization (MWO). This Organization allowed control over every national weather agency. It worked two ways. The national weather agency nominated members of the IPCC who produced the Summary for Policymakers that they then pushed on the politicians. Bureaucratic scientists challenged politicians who dared to ask questions. In most countries, these bureaucrats also controlled all the funding for climate research and directed it to research to perpetuate the fable produced by the IPCC. It was no coincidence that the SPM was released just prior to annual monetary decisions for funding the IPCC and climate change.

Three national weather agencies were central to controlling the IPCC and the research they produced. The United Kingdom Meteorological Office (UKMO) was most important because the Director, Sir John Houghton became the first co-chair of the IPCC. The UKMO's working connection ensured CRU people appointments to the IPCC. Probably because of Maurice Strong's

Chapter Thirteen

Canadian connections Environment Canada participated from the start. Bureaucrat Gordon McBean, Assistant Deputy Minister at the agency chaired the formation meeting of the IPCC at Villach Austria in 1985. The third agency was the National Oceanic and Atmospheric Administration (NOAA), which became more directly involved in the important 2001 IPCC Report under the Co-chair of NOAA employee Susan Solomon. Ms. Solomon was critically involved previously in the CFC ozone issue.

It is remarkable how the IPCC survived the leaked emails and the "hockey stick" fiasco. Despite all the evidence of corruption and bad science the PR people with the help of government agencies and universities were able to perpetuate the deception. Most people still think CO_2 is a problem, but more on that shortly.

After 1998 CO_2 levels increased, but despite their efforts temperatures leveled and declined. IPCC projections were wrong. According to their hypothesis this could not happen. They failed as renowned physicist Richard Feynman explains:

> In general, we look for a new law by the following process: First we guess it; then we compute the consequences of the guess to see what would be implied if this law that we guessed is right; then we compare the result of the computation to nature, with experiment or experience, compare it directly with observation, to see if it works. If it disagrees with experiment, it is wrong. In that simple statement is the key to science. It does not make any difference how beautiful your guess is, it does not make any difference how smart you are, who made the guess, or what his name is—if it disagrees with experiment, it is wrong.

IPCC acknowledged the problem in their latest Report (AR5) released in September 2013, but ignored it, as usual. Some claim

AR5 is new because it considers a previously ignored solar mechanism, but the conclusion is more incorrect than in AR4. AR5 concludes:

> *Globally, CO_2 is the strongest driver of climate change compared to other changes in the atmospheric composition, and changes in surface conditions. Its relative contribution has further increased since the 1980s and by far outweighs the contributions from natural drivers.*

This is only true because they did not consider most *natural drivers*. They created unreal explanations, ignored contradictory evidence, used computer model generated data as real data in other computer models and used theoretical ideas as real. They made it up as they went along. They also moved the goalposts. Figure 36 shows the global temperature over the last several years. Starting in 1998 global temperatures began to level then decline. (Figure 36) Since then the declined has continued in a natural pattern related to declining solar activity. The trouble was CO_2 levels continued to increase in direct contradiction to their hypothesis. According to the IPCC 90+ certainty claim, this cannot happen. It was a perfect example of T. H Huxley's observation, "The great tragedy of science—the slaying of a beautiful hypothesis by an ugly fact." Proper science would go back and reconsider the hypothesis. Instead, the alarmists simply moved the goalposts by changing the terminology. What was global warming caused by CO_2 became climate change.

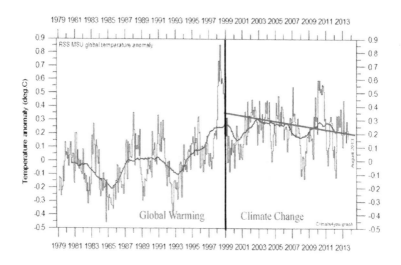

Figure 36

The Switch from Global Warming to Climate Change

The IPCC had already switched from predictions to scenarios. These included low, medium and high temperature projections primarily determined by increasing industrial growth and the associate production of CO_2. This correlation had appeared somewhat related in earlier projections, but now it was increasingly incorrect. Figure 37 shows the three projections compared to the actual temperature over the same period. [231]

---

[231] http://clivebest.com/blog/?p=2208

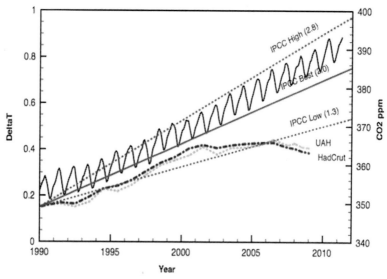

Figure 37

CO₂ Projections

When you plot the IPCC data for CO$_2$ levels over the same period it becomes obvious the extent of the influence on their projections:

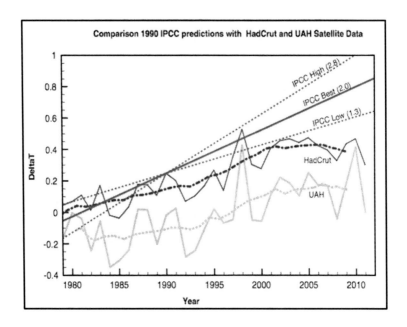

Figure 38[232]

CO_2 Projections Compared to Satellite Data

Sunspot activity correlates far better with the temperature than CO_2, but the IPCC do not include the scientific evidence of the relationship between sunspots and global temperatures. History and the best scientific evidence indicate continued decline in sunspot activity and therefore lower global temperatures. The problem is governments are preparing for warming.

Now we must deal with the economic damage created by corrupted political science because the IPCC achieved its goal of demonizing CO_2. Amazingly, it was done in full view. It was achieved because most people—including skeptics and deniers—still don't understand even the most basic elements of the science. All the efforts of the so-called skeptics or deniers had little effect.

---

[232] http://clivebest.com/blog/?p=2208

It was achieved because most people were directed to look at one piece of a very extensive and complex system.

The IPCC concluded that:

> *Another unusual aspect of recent climate change is its cause: past climate changes were natural in origin (see FAQ 6.1), whereas most of the warming of the past 50 years is attributable to human activities...*

...without raising red flags. How can they make such a claim when the variation of just one minor variable, albedo (the percentage of sunlight reflected back to space by the colour of the Planet), alone exceeds the entire change due to human CO_2?

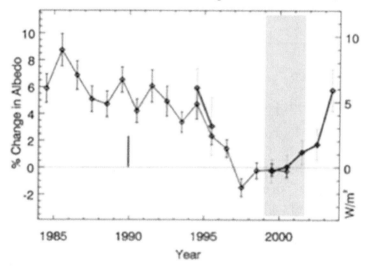

Figure 39

Global Albedo Change 1984-2004. Error Bars Due to Seasonal Variability of 15-20%[233]

---

[233] Source: http://www.skepticalscience.com/earth-albedo-effect.htm

Or consider Dr. Rao's findings about cosmic rays that the IPCC don't even consider that the contribution of decreasing cosmic ray activity to climate change is almost 40 percent, in a paper accepted for publication in *Current Science*, the preeminent Indian science journal. The IPCC model, on the other hand, says that the contribution of carbon emissions is over 90 percent.

IPCC's success in achieving their goal is because of the deliberate process and methods put in place to direct and control their work. It is also due to the generalist nature of climatology in an age of specialization.[234] I spoke of this in my presentation to the First Heartland Conference in New York. The result is the IPCC was able to focus on individual components and present them out of context. They were able to put and keep the focus on CO_2 so that most people still don't know it is less than 4 percent of the greenhouse gases or even that water vapor is the most abundant and important greenhouse gas.

It is true that concern about warming has diminished, overridden by economic issues and failed predictions, especially in Europe. The major problem was the switch from the term global warming to climate change. It was further amplified by another PR term, "climate disruptions", used by John Holdren, Obama's science czar.[235] This reinforced the practice of implying or stating that current natural climate change events are unnatural.

The world is still focused on CO_2, although most now incorrectly refer to it as carbon—including the U.S. President. Most governments have policy for CO_2 reduction; many impose direct and indirect taxes and other forms of legislation restricting CO_2 production. They reinforced this policy with subsidies to alternate energies because they produce less CO_2. This almost single focus was the objective of the IPCC. Despite a steady flow of scientific challenges to its role in weather and climate, coupled

---

[234] http://drtimball.com/2011/generalist/
[235] http://drtimball.com/2009/john-holdren/

with exposure of corrupt activities, it continues to dominate global energy and environmental policies. Private industry has seized the opportunities created by the focus to produce and promote products and to improve their public image. Consider car manufacturers promoting low CO_2 output as a sales feature. In a promotion for new vehicles Mercedes says, *Mercedes-Benz products to drop CO_2 emissions below 140g/km by 2012 —and two new models will lead the charge.*[236]

The meaningless term "Carbon Footprint" is now part of common language.

The narrow prism was deliberately chosen and the science, bureaucratic structure and propaganda designed to 'prove' CO_2, particularly the human portion, was causing catastrophic warming and then climate change. They kept almost everyone focused on the CO_2 because few understand climatology and they were outmaneuvered by the propaganda. On June 9th, 2011 the *Indian Express* reported:

> Declaring that "science is politics in climate change; climate science is politics", Union Environment Minister Jairam Ramesh on Wednesday urged Indian scientists to undertake more studies and publish them vigorously to prevent India and other developing countries from being led by our noses by Western (climate) scientists who have less of a scientific agenda and more of a political agenda.

This statement is only necessary because despite disclosure of leaked emails, the malpractice of the 'hockey stick', failure of 'predictions', and many other serious problems, the IPCC has achieved its goal.

The Indian Minister's direction to the scientists' borders on

---

[236] http://www.motoring.com.au/news/2010/mercedes-benz/paris-show-low-carbon-mercedes-cls-s-class-21874

the type of political interference he complains about. It is why government must be removed from climate research completely. Government agencies were used to create and perpetuate corrupted science through the IPCC, so it must be disbanded. The WMO and all government weather agencies must be restricted to data collection and dissemination—no research. Bureaucrats doing research almost guarantee it will be political. Once the Weather bureaucracies adopted the IPCC Reports as gospel, they worked to block research that disproved the AGW hypothesis, thus defeating the scientific method. Weather and climate research funding can come from government—but only through arms-length agencies.

John Maynard Keynes said:

> *When the facts change, I change my mind. What do you do, sir?*

Keynes comment is especially applicable to science. It is the duty of all scientists to question a new hypothesis and find facts that might change minds. The scientific method requires that you try to disprove the hypothesis. The IPCC hypothesis proposed that an increase in CO_2 would result in global warming and it would increase because of human additions from a developed world. They excluded and modified data and facts instead of letting it reject or reset their hypothesis. The constant and persistent effort by the IPCC was to prove the hypothesis.

All these practices continue as evidenced by a leak of the draft IPCC Report known as AR5.[237] One group tried to re-establish the claim that the Antarctic ice core record showed CO_2 increase preceded temperature increase. Another, including one of the same authors, tried to produce a "hockey stick" type graph

---

[237] http://wattsupwiththat.com/2012/12/13/ipcc-ar5-draft-leaked-contains-game-changing-admission-of-enhanced-solar-forcing/

to show that 20th century warming was unprecedented. It was achieved using the similar technique of grafting a doctored modern temperature record on to a proxy record of a few samples around the world. The latter study was released in time for inclusion in AR5, but not leaving sufficient time for peer reviewed rebuttals.

The greatest deception continues for a variety of reasons. They are the same reasons why a diversity of people became involved:

- They were part of the cabal who chose climate and environment as vehicles for their political agenda.
- They were academics attracted by the funding offered in massive amounts.
- They were academics seeking the funding, but also with political sympathies with the cabal's objectives.
- They were bureaucrats pulled in by the national organizational structure chosen through the UN World Meteorological Organization (WMO).
- They were bureaucrats with political sympathies with the cabal objectives.
- They were politicians who saw an opportunity to "be green."
- They were businesses that saw an opportunity for a profitable business guaranteed by government policy and funding.
- They were people who simply saw a business opportunity.
- They were global environmental groups who supported the political objectives of blaming humans for the world's ills.
- They were Non-Government Organizations (NGOs). The term coined by the UN in 1945, but reconstituted by Maurice Strong for the Rio 1992 conference and comprising organizations not part of a government or conventional for-profit businesses.

- They were politicians who saw an opportunity for more taxation.

The Damage and the Irony—Overpopulation

The underlying theme of the Club of Rome and Agenda 21 was a Neo-Malthusian belief that the world is overpopulated. At the population conference in Cairo Al Gore said:

> *No single solution will be sufficient by itself to produce the patterns of change so badly needed.*

The comment is true on its face, but Gore's solutions have already failed and will do further damage if continued. Ironically, the solution to overpopulation is already known and proved. The problem for Gore the Club of Rome and Agenda 21 is it is in direct opposition to their solutions. Their predictions were wrong but more important another pattern had emerged. It is known as the Demographic Transition Model (Figure 40).[238] It shows that populations decline with economic development—the very thing the entire Strong orchestrated agenda was designed to oppose.

---

[238]

http://geography.about.com/od/culturalgeography/a/demotransition.htm

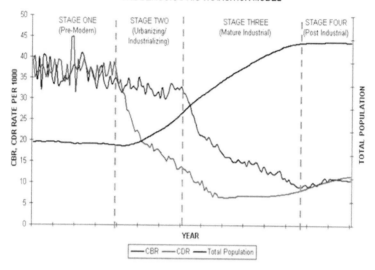

THE DEMOGRAPHIC TRANSITION MODEL

Figure 40

The Demographic Transition Model

It shows, and all the statistics confirm, that populations decline as nations industrialize and the economy grows. Ironically, the problem in most developed countries is too rapid a decline in population and insufficient young people to support the massively expensive social programs for the elderly. Many of them offset the population decline with immigration. Generally supporters of the Gore/Holdren/Ehrlich/Agenda 21 approach favor immigration, but in reality it only papers over the problems while creating others. Most emigrants are economic migrants seeking better economies. If the economies were better where they were born or live they likely would not migrate.

Alternate Energies and Green Economies

The false demonization of CO_2 was to reduce industry by reducing fossil fuels. It was accompanied by plans to replace fossil fuels with alternate energies that would create new green jobs and economies. It failed everywhere it was tried. Many countries and regions are already reversing their energy plans. For example, Europe is going back to coal.[239] Figure 41[240] shows a view in Britain as power bills soar.

BRITAIN IS BETTER OFF WITHOUT THE CLIMATE CHANGE ACT

Figure 41

Green Isn't Working

Rather than discuss all the countries and their well-documented problems there is a region that typifies the problems when the green agenda is implemented. It is particularly appropriate

---

[239] http://www.thegwpf.org/europe-cutting-green-energy-coal-comeback/

[240] The Global Warming Policy Foundation; Newsletter CCNet 14/10/13

because Maurice Strong, father of the UNEP and IPCC, was the one who instituted it.

Citizens of the Canadian Province of Ontario are paying for the green energy agendas created by Maurice Strong when he was head of Ontario Hydro, the public utility that controls all grid power. He was subsequently aided and abetted by environmentalist David Suzuki and Provincial Premier Dalton McGuinty who pushed Strong's disastrous policies. It will take years to rebuild adequate facilities. They ignored the science, but that is the pattern of politicians from former U.S. Senator Timothy Wirth in 1993 who said:

> We've got to ride the global warming issue. Even if the theory of global warming is wrong, we will be doing the right thing.

...to Canadian Environment Minister Christine Stewart who said:

> No matter if the science of global warming is all phony... climate change provides the greatest opportunity to bring about justice and equality in the world.

Strong began Ontario Hydro's problems when appointed Chairman by NDP Premier Bob Rae in 1992. A 1997 article titled *Maurice Strong: The new guy in your future* says:

> Maurice Strong has demonstrated an uncanny ability to manipulate people, institutions, governments, and events to achieve the outcome he desires...The fox has been given the assignment, and all the tools necessary, to

repair the henhouse to his liking.[241]

This applies to his UN role, but also applies to his Ontario Hydro role.

One report says:

> *Within no time of his arrival, he firmly redirected and re-structured Ontario Hydro. At the time, Ontario Hydro was hell-bent on building many more nuclear reactors, despite dropping demand and rising prices. Maurice Strong grabbed the Corporation by the scruff of the neck, reduced the workforce by one third, stopped the nuclear expansion plans, cut capital expenditures, froze the price of electricity, pushed for sustainable development, made business units more accountable.*
>
> *Sounded good, but it was a path to inadequate supply.*[242]

Key is the phrase he, *pushed for sustainable development*. In the same year, 1992, Strong, in the keynote speech at the Rio Earth Summit he organized, said:

> *Current lifestyles and consumption patterns of the affluent middle class—involving high meat intake, the use of fossil fuels, electrical appliances, home and workplace air-conditioning, and suburban housing—are not sustainable.*

---

[241]

http://www.citizenreviewonline.org/august_2002/maurice_strong.htm

[242] http://www.hydro-gate.com/newpage41.htm

He'd already created mechanisms to eliminate fossil fuels and bring about reduction and destruction of western economies, including Ontario. It's summarized in his speculation:

> *Isn't the only hope for the planet that the industrialized civilizations collapse? Isn't it our responsibility to bring that about?*

How do you cause industrial civilizations to collapse? You cut off their energy supply. Fossil fuels drive the industrial economies and CO_2 is a byproduct. Show it is causing irreparable climate damage and you can demand alternative energy replacements. Strong achieved this with the IPCC and at Ontario Hydro. He used the narrow definition of climate change created by the United Nations Framework Convention on Climate Change (UNFCCC), as only human caused changes. Trouble is this is impossible if you don't know the amount and cause of natural change. Ontario is in trouble because it switched to alternate energies. It is in further trouble because the world is cooling naturally, not warming as Strong's agenda assumed.

IPCC created the science to prove human CO_2 was the problem and the politics to claim failure to act guarantees catastrophe. Strong controlled who participated through the bureaucratic members of the World Meteorological Organization (WMO). An Assistant Deputy Minister of Environment Canada (EC), who subsequently controlled most Canadian climate research funding, chaired the IPCC formation meeting in 1985. Using Weather Departments gave bureaucrats ascendancy over politicians.

CO_2 is not causing warming or climate change. There is no scientific need to replace fossil fuels. Replacing them with alternative energies compounds the problems. A U.S. Senate report notes:

Comparisons of wind, solar, nuclear, natural gas and coal sources of power coming on line by 2015 show that solar power will be 173% more expensive per unit of energy delivered than traditional coal power, 140% more than nuclear power and natural gas and 92% more expensive than wind power. Wind power is 42% more expensive than nuclear and natural gas power.... Wind and solar's "capacity factor" or availability to supply power is around 33%, which means 67% of the time wind and solar cannot supply power and must be supplemented by a traditional energy source such as nuclear, natural gas or coal.[243]

Wind turbulence restricts the number of turbines to 5 to 8 turbines per 2.6 square kilometers. With average wind speeds of 24 kph it needs 8,500 turbines covering 2590 square kilometers to produce the power of a 1000 MW conventional station. To put this in perspective Ontario closed two 1000 MW plants in 2011—the Lambton and the Nanticoke coal fired plants. You need 5,180 sq kms of land to replace them with wind power. Besides the land, you still need coal-fired plants running at almost 100 percent for back up. But the Ontario situation only relates to coal, nuclear and wind. There are similar limitations on all alternate energies.

We're in this predicament because of exploitation by politicians and environmental groups who deliberately ignore scientific evidence and corruption in climate science. Options were dramatically reduced by campaigns of fear against nuclear power creating legislation so that it now takes up to 14 years to

[243]

http://www.naturalclimatechange.us/natural%20climate%20change-%20adds%2012-3-10/Quotes%20on%20man-caused%20global%20warming.pdf

construct a nuclear power plant. [244]

Capabilities of alternative energies were misrepresented and real costs grossly distorted by subsidies. There are so many subsidies at so many different stages that it is probably impossible to do accurate and useable cost/benefit analyses. One definition says:

> *Alternative energy is an umbrella term that refers to any source of useable energy intended to replace fuel sources without the undesired consequences of the replaced fuels.*[245]

If this were true what people consider alternative energies would qualify as "replaced fuels." It is a cute academic definition, but the reality is the only fuels considered "undesirable" are those that produce CO_2. This is because of the deceptive work of the IPCC. People forget that their predictions of future temperatures are based on continued and increasing demand for electricity—business as usual. They cannot anticipate technological innovations. For example, though expensive at present, Light Emitting Diode (LED) white light will dramatically reduce power requirements.

Some define renewable energies as the only acceptable alternative energies, but they have severe limitations. The focus is diverted away from the real power production problems and potential resolutions. Even President Obama concedes they're not a short-term solution, but they're not a long-term solution either. Major energies touted are wind and solar, but in 2007 they provided only 6 percent of alternative U.S. production, which is

---

244

http://nuclearinfo.net/Nuclearpower/WebHomeCostOfNuclearPower

[245] http://www3.jjc.edu/ftp/wdc11/webtechbinary11bi/index.html

just 7 percent of total U.S. production (Figure 42). Percentages have changed slightly but are still insignificant and limitations continue.

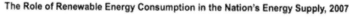

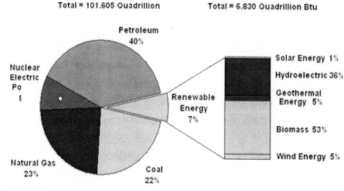

The Role of Renewable Energy Consumption in the Nation's Energy Supply

Figure 42[246]

Except for petroleum to drive vehicles the common denominator of any energy source is to produce electricity. The overarching need is for a continuous supply of energy. Solar and wind are not continuous, so they must have backup sources instantly available and coal, oil, or nuclear are the only options. Countries that have attempted wind power experience an increase in CO_2 production. An article titled *"Wind power is a complete disaster"* reports the German experience:

---

[246] http://seekingalpha.com/article/82445-keeping-alternate-energy-in-perspective

Germany's CO_2 emissions haven't been reduced by even a single gram—and additional coal and gas-fired plants have been constructed to ensure reliable delivery.[247]

Denmark has the highest percentage of wind power and their experience is telling. As the National Post article reports:

Its electricity generation costs are the highest in Europe (15¢/kwh compared to Ontario's current rate of about 6¢). Niels Gram of the Danish Federation of Industries says, "windmills are a mistake and economically make no sense." Aase Madsen , the Chair of Energy Policy in the Danish Parliament, calls it "a terribly expensive disaster."

Because the wind can drop or surge suddenly it puts stress that can overload the grid so wind power is generally limited to 12 percent of the total supply. Other problems include the surge demand placed on the grid when the wind drops off, or the addition of surplus power when the electrical demand drops and wind power is still being added. A report from Britain tells of wind farms being paid to shut down turbines to prevent this problem.[248] These are economic realities, but add in the number of birds killed, the blight on the landscape, and the cost of transmission from remote locations and it is not an alternative.

Solar power is no better. Spain has paid the price and has moved to stop the bleeding.

Spain is lancing an 18 billion-euro ($24 billion) investment bubble in solar energy that has boosted public

---

[247] http://www.freerepublic.com/focus/f-news/2408971/posts
[248]

http://www.telegraph.co.uk/earth/energy/windpower/7840035/Firms-paid-to-shut-down-wind-farms-when-the-wind-is-blowing.html

liabilities choking off new projects as it works to cut power prices and insulate itself from Greece's debt crisis.[249]

Loss of supply is slightly more predictable because of known hours of daylight. However, these are less than half the day in winter for most of the world. The unknown factor is cloud cover. You reduce this by going to desert regions but then there is windblown sand damage, as well as vast arrays despoiling landscapes and ecologies.

We've already experienced limitations of biofuels triggered by government subsidies. In that case it was simply U.S. agricultural land diverted. As one review notes:

Switching to biodiesel on a large scale requires considerable use of our arable area. Even modest usages of biodiesel would consume almost all cropland in some countries in Europe![250]

More CO_2 and nitrogen oxides are produced than from fossil fuels, but they divert from this reality by presenting a net figure achieved by subtracting CO_2 used to grow the plant. Biodiesel has lower fuel efficiency than petro-diesel. Low temperatures are a serious limit for all diesels but worse for biodiesels.

Geothermal has potential but is limited in location and usually far from where it is needed. The same is true for hydroelectric and tidal power. If the world really wants to solve the energy problems a prize for a method of reducing line loss,

---

[249] http://www.bloomberg.com/news/2010-04-29/spain-pricks-solar-bubble-and-loses-investors-to-avoid-greek-style-crisis.html

[250]

http://www.billdoll.com/dir/science/energy/q/biofuels_make_cut.html

and another for a method to effectively store electricity would be better than alternate energies.

Promoting energy policies based on falsified science and alternative energies that don't work is unacceptable. When they've been tried and already failed as in Ontario, it is incredible anyone would continue to promote them. The people will pay the price as they have already.

Final Stages of an Exploited Paradigm Shift

The sad truth is none of the energy and economic policies triggered by the demonization of CO_2 were necessary. The price paid even to date is incalculable and it will go on costing for too many people, especially scientists and governments, who built their entire careers around the falsehoods. Environmentalism was an essential paradigm shift and as usual the majority saw the general sense and benefits but were unsure about the limits. The situation was complicated as a small group took environmentalism and the subset of climate to push through a political agenda. Extremists define the limits for the majority by behaving or promoting ideas and actions that most can define as detrimental. In the case of environmentalism and climate those with the political agenda became the extremists.

Because they applied politics to science they perverted the scientific method by proving their hypothesis to predetermine the result. They bullied, marginalized, and silenced skeptics for a while, but natural events contradicted their predictions. Instead of acknowledging the error of their science they orchestrated a desperate counterattack. There are two major dangers of what was done. 1. The credibility of science, especially climatology will suffer badly. 2. Environmental issues will suffer because people will say you lied to us before why should we believe you now. There are a multitude of minor damages that people become aware of as the switch to alternative energies, green economies and green jobs continue.

Hopefully, this book has exposed the greatest deception in science and human history; what was done and how it was done. Problems are only a problem if you are unaware of them. Once the problem is defined it is possible to define solutions. It is time to demand accountability and set up processes and procedures to prevent such global deceptions.

As Samuel Johnson said:

> *Hope is itself a species of happiness, and perhaps, the chief happiness which this world affords.*

Other books on global warming from Stairway Press.
Available anywhere on Earth.

www.StairwayPress.com

Other books on global warming from Stairway Press.
Available anywhere on Earth.

www.StairwayPress.com

CPSIA information can be obtained at www.ICGtesting.com
Printed in the USA
LVOW06s1211300514

387793LV00034B/569/P